The Science and Development of Transport—ZIRP 2021

Marjana Petrović · Luka Novačko · Diana Božić ·
Tomislav Rožić
Editors

The Science and Development of Transport—ZIRP 2021

Springer

Editors
Marjana Petrović
Department of Railway Transport
Faculty of Transport and Traffic Sciences
University of Zagreb
Zagreb, Croatia

Luka Novačko
Department of Road Transport
Faculty of Transport and Traffic Sciences
University of Zagreb
Zagreb, Croatia

Diana Božić
Department of Transport Logistics
Faculty of Transport and Traffic Sciences
University of Zagreb
Zagreb, Croatia

Tomislav Rožić
Department of Transport Logistics
Faculty of Transport and Traffic Sciences
University of Zagreb
Zagreb, Croatia

ISBN 978-3-030-97527-2 ISBN 978-3-030-97528-9 (eBook)
https://doi.org/10.1007/978-3-030-97528-9

This Springer imprint is published by the registered company Springer Nature Switzerland AG
The registered company address is: Gewerbestrasse 11, 6330 Cham, Switzerland

Preface

This monograph presents the original scientific manuscripts submitted for publication at the International Conference - The Science and Development of Transport (ZIRP 2021), organized by the University of Zagreb, Faculty of Transport and Traffic Sciences, Zagreb, Croatia. The co-organizers of the ZIRP 2021 conference are Wrocław University of Science and Technology, University of Bremen, University of Zagreb Faculty of Mechanical Engineering and Naval Architecture, Croatian Chamber of Transport Engineers, and AMAC-FSC. The Conference was a hybrid event, held virtually and in person, in Šibenik, Croatia, on 30 September and 1 October 2021.

The Conference covered many relevant topics, and the participants discussed issues such as sustainable urban mobility and logistics, safety and policy, data science, process automation, inventory forecasting, improving competitiveness in the transport and logistics services market, and raising customer satisfaction. The Conference's main aim is to inspire innovation in the transport industry by allowing scientists and practitioners to work together.

The editorial board has chosen 14 papers covering relevant issues in today's transport and logistics industry for the monograph. The monograph will be of interest to experienced researchers and professionals as well as Ph.D. students in the field of transport and logistics. We would like to express our gratitude to the authors that contributed to this monograph.

In the chapter "Assessing Smart City Initiatives: A Case Study of Croatian Municipalities", the authors analyzed and compared five Croatian municipalities based on the selected CITYkeys indicators. The conducted research revealed a correlation between the number of indicators that were met and the level of development of the county in which the municipality is located.

Zagreb saw an increase in bicycle traffic in recent years. The authors of the chapter "Geospatial Analysis of Bicycle Infrastructure in the Function of Traffic Safety: The Case of the City of Zagreb" have conducted a much-needed analysis of the existing cycling routes. In line with the safety aspect of the analyzed infrastructure, the authors determine the volume of traffic flow in specific locations with a higher chance for a conflict between bicycles and other traffic flows.

The research presented in the chapter "University Infrastructure, Sustainable, and Equitable Accessibility? Study Case: Manizales—Villamaría (Colombia) Metropolitan Area" evaluated the geographical accessibility of educational facilities by establishing whether a private vehicle or public transport is used. The study employed the method of integral average accessibility, using the average travel time needed to reach the facilities as a comparative measure. Additionally, an analysis of population coverage and socioeconomic equity was performed.

The chapter "Functioning Paid Parking Zone Before and During COVID-19 Pandemic—Case Study on the Example of a City in Poland" examines the functioning of PPZ in Chorzów (Poland) before and during the COVID-19 pandemic. The analysis was carried out based on the number of purchased parking tickets in particular months, days of the week, and periods of the day in 2019 and 2020.

The chapter "Speed Transition Matrix Feature Extraction for Traffic State Estimation Using Machine Learning Algorithms" identifies the key features for traffic state estimation extracted from the speed transition matrix that enable the simplification of the learning process and the interpretability of the obtained results. The proposed features are then used to estimate the traffic state for the most crucial road segments in the City of Zagreb, Croatia.

The chapter "Are We Ready for Autonomous Vessels?" addresses the phenomenon of autonomous vessels inclusion in Croatia legislation. The author analyzes the current state of legislation regarding autonomous vessels worldwide and provides recommendations for possible solutions regarding liability issues concerning this new kind of vessel.

Autonomous, green vessels in urban environments have the potential to become long-term and sustainable solutions that, apart from the revitalization of the urban water transport, will enhance the economic, ecological, safety, and even tourist aspects of the cities. The chapter "Possibilities of Using Autonomous Green Vessels for Passenger Transport in Urban Environments" gives an outline and analysis of the current state of the sea/water transport in some of the most popular cities of the world and an overview of the state-of-the-art projects and vessels suitable for urban environments.

The chapter "State and Projection of the Port Traffic in the Port of Split: Post–COVID Era" analyses the total port traffic in the Port of Split comprising both the passenger port basin, known as the City port, and cargo terminals situated in the northern part of the city, dubbed the North port. In addition to the presented statistical analysis, the final part of the chapter provides an assessment of the ship traffic recovery in the Port of Split based on linear trend projection.

There has been growing interest in the logistics industry within the context of improving environmental sustainability in developing countries. The topic is addressed in the chapter "Assessing Barriers of Low-Carbon Emission Operations in Logistics Service Providers in Vietnam". In-depth interviews and a survey involving 159 logistics service companies were conducted to collect data about eight groups of barriers to reducing carbon emission in practice.

Since waste management is one of the crucial economic branches of the future, both from the aspect of circular economy and the aspect of environmental protection, all the components, including transport, represent challenges in the everyday requirements for an acceptable transport solution in the freight transport. The chapter "Development of Innovative Rail Wagons for Transport of Municipal Waste "EKO–VAKO"" discusses the issue of transporting municipal waste from the place of origin to the recycling centers by using rail transport.

The research presented in the chapter "Assessment of the Impact of Technological Development and Organizational Complexity in Air Transport" addresses the elements that can affect the organization-technology relationship whose impact can be determined (from a safety perspective) by the influence of the human factor. After performing a critical review of the literature regarding organizational safety and the current stage of technological development, the chapter approaches the subject of organizational behavior managed toward improved adaptability and safety risk minimization.

The chapter "Strategic Settings for the Development of Franjo Tuđman Airport Cargo City as a Regional Center" analyses the capacities of the existing infrastructure and estimates the future demand for air cargo transportation in the airport's catchment area. The research aims to determine the future needs and possibilities of the airport regarding current capacities and existing infrastructure in terms of increasing business interest in air cargo transport.

The chapter "CFD Analysis of F-16 Wing Airfoil Aerodynamics in Supersonic Flow" shows how a simple simulation-based analysis is better at predicting aerodynamic characteristics of a wing with changing geometry. The chapter also promotes wing morphing technologies as they increase aerodynamic characteristics when the geometry changes, reducing fuel consumption and noise.

Military flying operations performed in flexible airspace structures impact commercial air traffic. Therefore it is necessary to analyze the planning process and develop a prediction tool. The chapter "Prediction Tool for Military Flying Operations Planning" addresses this complex issue.

Zagreb, Croatia

Marjana Petrović, Ph.D.
Luka Novačko, Ph.D.
Diana Božić, Ph.D.
Tomislav Rožić, Ph.D.

Contents

Assessing Smart City Initiatives: A Case Study of Croatian Municipalities 1
Maja Mutavdžija, Majda Kodžaga, Predrag Brlek, and Krešimir Buntak

Geospatial Analysis of Bicycle Infrastructure in the Function of Traffic Safety: The Case of the City of Zagreb 15
Dario Oršić, Mario Ćosić, Ljupko Šimunović, and Matija Sikirić

University Infrastructure, Sustainable, and Equitable Accessibility? Study Case: Manizales–Villamaría (Colombia) Metropolitan Area 33
Juan M. Gómez, Carlos A. Moncada, and Diego A. Escobar

Functioning Paid Parking Zone Before and During COVID-19 Pandemic—Case Study on the Example of a City in Poland 47
Agata Kurek and Elżbieta Macioszek

Speed Transition Matrix Feature Extraction for Traffic State Estimation Using Machine Learning Algorithms 61
Leo Tišljarić, Filip Ribić, Željko Majstorović, and Tonči Carić

Are We Ready for Autonomous Vessels? 75
Iva Savić

Possibilities of Using Autonomous Green Vessels for Passenger Transport in Urban Environments 91
Mirko Čorić and Mladen Nikšić

State and Projection of the Port Traffic in the Port of Split: Post-COVID Era 111
Zvonimir Lušić, Luka Vukić, Danijel Pušić, and Ivica Skoko

Assessing Barriers of Low-Carbon Emission Operations in Logistics Service Providers in Vietnam ... 133
Thi Thu Huong Tran, Van Kiem Pham, Phuong Thao Vu, Thu Trang Pham, and Khac Huy Nguyen

Development of Innovative Rail Wagons for Transport of Municipal Waste "EKO-VAKO" ... 155
Dražen Vidović, Mladen Nikšić, and Monika Šušak

Assessment of the Impact of Technological Development and Organizational Complexity in Air Transport ... 167
Valentin Marian Iordache, Sorin Eugen Zaharia, and Casandra Venera Pietreanu

Strategic Settings for the Development of Franjo Tuđman Airport Cargo City as a Regional Centre ... 181
Jure Šarić, Andrija Vidović, Igor Štimac, and Tomislav Mihetec

CFD Analysis of F-16 Wing Airfoil Aerodynamics in Supersonic Flow ... 197
Leo Matak and Karolina Krajček Nikolić

Prediction Tool for Military Flying Operations Planning ... 211
Nikola Mostarac, Darko Šumanovac, and Doris Novak

Assessing Smart City Initiatives: A Case Study of Croatian Municipalities

Maja Mutavdžija, Majda Kodžaga, Predrag Brlek, and Krešimir Buntak

Abstract The smart city concept is increasingly represented in Europe, where many cities try to achieve as high a level of smartness as possible, through different smart city initiatives. Smart city initiatives are not only limited to urban areas, but also to wider and narrower geographical areas. Such initiatives are also implemented in rural areas and municipalities, better known as smart villages and smart municipalities. Given the possibility of implementing various smart solutions at multiple levels, there is a need to evaluate such solutions and initiatives. Within the HORIZON 2020 programme, a set of indicators for evaluating smart initiatives was developed through CITYkeys methodology. Based on selected indicators from this methodology, five municipalities on the territory of the Republic of Croatia were analysed and compared. The conducted research shows that the municipality that meets the most indicators, but also has quite high results, comes from Varazdin County, while the lowest results are identified in the municipality positioned in Koprivnica-Krizevci County and Virovitica-Podravina County, which is also one of the least developed self-governments unit in the Republic of Croatia.

Keywords Smart cities · Smart municipalities · CITYkeys · Mobility · Quality of life

1 Introduction

The development of the concept of smart cities begins with the process of urbanization. Urbanization is defined as a process which transforms traditional settlements into urban environments under the influence of various achievements, the most significant of which is the technological one. Accordingly, there are challenges related to development and life in a created urban environment [1]. The concept of smart cities introduces a number of new initiatives, not only applicable to urban areas, but also to smaller areas such as municipalities and rural areas. For such an area to achieve

M. Mutavdžija (✉) · M. Kodžaga · P. Brlek · K. Buntak
Department for Logistics and Sustainable Mobility, University North, Trg dr. Žarka Dolinara 1, 48000 Koprivnica, Croatia
e-mail: mamutavdzija@unin.hr

M. Petrović et al. (eds.), *The Science and Development of Transport—ZIRP 2021*,
https://doi.org/10.1007/978-3-030-97528-9_1

progress and to be considered smart, it should be able to pool all its resources and act effectively to meet the set objectives [2]. Cities, municipalities and rural areas around the world have started looking for solutions that enable transport connections, mixed land use and high-quality urban services with long-term positive effects on the economy. For example, one of the key elements for city's growth is efficient public transport that meets economic needs. Many new approaches concerning urban services are based on the exploitation of new technologies [3]. Smart cities' philosophy is to look at challenges as opportunities and take advantage of other trends such as digitalization [4]. But it is also important to highlight several barriers that limit the potential of innovative smart technologies, such as high technological risk, user security and privacy, financial and business difficulties, uncertain return on investment or regulatory difficulties, and lack of cooperation [4]. When designing smart city initiative, it is necessary to look at all aspects, namely all advantages and opportunities, as well as all potential risks.

Given the fact that many cities, municipalities and villages implement smart initiatives, various initiatives need to be evaluated. Existing research on implemented smart initiatives is exclusively related to smart cities and specific projects and project proposals for the observed area. There are few studies in early stage related to the development of smart villages, but only few researches mention municipalities and similar spatial units. In the territory of the Republic of Croatia, municipalities are defined as the lowest level political–territorial units, and are defined by law as units of local self-government. As such, municipalities have the opportunity to apply for funds to finance projects in the field of smart cities, and also implement numerous initiatives to improve life quality.

Different methodologies were suggested and developed to compare different initiatives within the concept of smart cities, and one of those methodologies is CITYkeys. This methodology contains a set of indicators categorized in several areas used for measuring and monitoring performance. The CITYkeys is financed by the European Union's HORIZON 2020 programme with assistance of cities and is used for collecting data for joint and transparent monitoring of smart city initiatives. The aim of this paper is to, through the methodology presented by the European Union, CITYkeys methodology, evaluate smart city initiatives implemented through projects in the territory of the Republic of Croatia, namely in selected Croatian municipalities. This paper aims to present the situation of individual Croatian municipalities in adapting to the postulates of sustainable development and the concept of smart cities, and compare the results to identify potential differences between selected municipalities. The research observes municipalities in the context of the counties where they are located, which have a significant impact on the performance of initiatives implemented in municipalities.

In addition to all the above, this paper consists of several basic chapters. The first part gives an overview of the literature related to the smart cities. The second part describes the use of the methodology, namely the CITYkeys methodology, which will be used for the analysis of selected municipalities according to selected indicators. The last part includes comparison of obtained results, discussion of conducted research and conclusion of the entire work.

2 Background

As for terminology, a smart city can be considered an intelligent city, a sustainable city, a digital city and a city aimed at achieving a high quality of life [5]. In this context, an intelligent city emphasizes its intellectual capital and creates synergy between each component of the city. The smart city concept is often seen through technology, as smart cities must use technological advances to improve their own services and improve the quality of life of inhabitants in the city. It is important to emphasize that smart cities are not based on technology, but need to use technology to make better choices and to manage the city. Due to the high level of technological advancement for a smart city, often used term is digital city, and this results from the digital transformation of numerous activities and services in the city. The technology itself in a smart city is also used to improve the quality of mobility, space and overall logistics. This is why smart cities are often called technology cities. What is crucial is that smart cities are achieving economic, ecological and social sustainability, and thus produce the highest possible quality of life for all citizens and visitors. A summary overview of key indicators for understanding the concept and definition of smart city is shown in Table 1.

Every smart city has several key components. The starting point is always a defined geographical area, i.e., the area where the city itself operates and where the population resides. On the other hand, the human component is one of the key components of smart city itself, encompassing all residents, visitors and decision makers. In addition to the human component, there is a connection between the public administration and the government of the city itself, i.e., persons who manage the city and make all relevant decisions for the development of the city. In addition, smart cities must implement certain technological solutions for more efficient and effective decision-making. From all this, a smart city has a spatial component, a human component, a control component and a technological component. Looking at the scope of a smart city, although the concept of smart city refers to an urban area, it is also possible to talk about smart villages and smart communities, which is a lower level than the smart city itself, because it covers a smaller geographical area. The concept of smart city, excluding urban areas, is also applicable to a higher level, such as regions, urban networks, the entire nation, and it is possible to speak

Table 1 Understanding smart cities [5]

	Explanation
Terminology	The concept of smart city can be compared with the term's intelligent city, digital city, sustainable city, technological city and city intended for quality of life
Components	Geographical area, technological infrastructure, services, management processes, population, government
Scope	Rural areas, municipalities, cities, regions, networks of cities, nations, global level
Goals	Ecological, economic and social sustainability, quality of life, participation in decision-making, knowledge and intellectual capital

about the global level. The smart city concept focuses on smart governance and is therefore applicable to all levels. Efforts are made through the smart cities concept and initiatives, to connect all elements of the city and to encourage the participation of citizens in making important decisions. Smart city initiatives are most often implemented through numerous projects that partially affect the improvement of the entire system.

3 Methodology

Smart city initiatives can be implemented at different levels. Considering that in the territory of the Republic of Croatia there are a total of 16 large cities out of 128 cities [6] covering more than 35,000 inhabitants and 428 municipalities, where the largest municipalities have up to 15,000 inhabitants, representing local self-government units, it is defined that research on smart initiatives should be carried out at the municipal level. For this purpose, four counties within three statistical regions of the second level were selected, and representative municipalities were selected within each county. CITYkeys methodology will be used to evaluate and compare smart city initiatives and projects. The aim of this methodology is to accelerate the introduction of smart city services and solutions in order to influence fundamental social challenges. The methodology aims to facilitate and enable all interested parties to learn from each other through projects, create trust and monitor progress through the set framework. Indicators developed through the CITYkeys methodology are in line with basic areas of sustainable development, which is an economic, environmental and social area. Therefore, the indicators are divided into groups as shown in Table 2.

Indicators for defined municipalities in the Republic of Croatia are selected according to the above categories. This methodology is also presented and suggested by Huovila et al. [8], Airaksinen et al. [9], Wendling et al. [10], Caird and Hallett [11] and others. The list of all indicators proposed through the CITYkeys methodology and selected for this research is presented in Table 3.

These indicators will be used to describe smart city initiatives in 5 municipalities on the territory of the Republic of Croatia; one from Koprivnica-Krizevci County, two from Virovitica-Podravina County, one from Varazdin County and one from Zadar

Table 2 Methodology indicators [7]

People	Planet	Prosperity	Governance	Propagation
Health, safety, access to services, education, diversity and social cohesion, quality of housing and the built environment	Energy and mitigation, materials, water and land, climate resilience, pollution and waste, ecosystem	Employment, equity, green economy, economic performance, innovation, attractiveness and competitiveness	Organization Community involvement Multi-level governance	Scalability, replicability

Table 3 Selected indicators from CITYkeys methodology [7]

Indicator	Unit of measurement	Definition
Access to basic health insurance	% people	Share of population with access to health services within a distance of 500 m
Encouraging healthy lifestyle	Scale from 1 to 5	The level of political efforts to foster a healthy lifestyle
Traffic accidents	#/1.000	Number of accidents per thousand inhabitants
Availability of public transport	% people	Number of persons with a public transport at a maximum distance of 500 m
Length of cycling track	% in km	% of cycling paths and lanes in relation to street length (excluding motorways)
Availability of high-speed internet	#/100	Number of users subscribed to fixed (wire) broadband network (per 100 inhabitants)
Environmental education	% of schools	% of schools that have environmental education in their programmes
Green area	ha/1.000	Quantity of green area in ha per 1000 inhabitants
Water consumption	(Average) no. of litres per person per day	Total water consumption per capita on a daily basis
Population density	$\#/km^2$	Number of persons per km^2
Municipal waste	(Average) weight of waste per capita in tonnes	Amount of municipal waste generated annually per capita
Percentage of waste recycling	% in tons	Percentage of waste to be recycled
Unemployment rate	% people	Percentage of unemployed labour force
Youth unemployment rate	% people	Percentage of unemployed young workforce
Share of green public procurement	% in HRK	Percentage of public procurement using the criteria of green public procurement in the implementation of public procurement procedures (total amount of green public procurement carried out/total amount of LGU) * 100 = % in HRK)

(continued)

Table 3 (continued)

Indicator	Unit of measurement	Definition
Gross domestic product	HRK/person	Amount of gross domestic product in the municipality/town per person
Newly registered operations	#/1.000	Number of newly registered jobs per 1000 inhabitants
Entrepreneurial incubators	#/1.000	Number of entrepreneurial incubators per 1000 inhabitants
Tourism intensity	#/1.000	Number of overnight stays per year per 1000 inhabitants
Availability of state data	Scale from 1 to 5	To what extent government data is available
Smart city policy	Scale from 1 to 5	To what extent the city/municipality has an incentive policy of a smart city
Expenditures by the municipality for a transition towards a smart city	HRK/person	Annual expenditures by the municipality for a transition towards a smart city

County. According to defined indicators, these municipalities will be compared in the next chapter. For privacy reasons, municipalities will not be named, but they will be categorized by the county in which they are positioned. We conducted interviews with the representatives of the municipalities and collected the requested data. The data was obtained exclusively from the competent persons. For indicators for which representatives do not have a defined value, the value is estimated, and this value is marked with the symbol *.

4 Results

Based on the conducted research and collected data, Table 4 is marked with poor, average and good values of the indicators. The values that are considered poor are marked in red, the average values are marked in yellow, while the good values of the required indicators are marked in green.

The assumed values (marked with *) are based on a review of the available literature, as well as the websites of the Croatian Employment Service, ministries or the police administration. In addition, telephone conversations were conducted with these institutions to obtain the assumed value of needed indicators.

When it comes to basic health insurance as a foundation for better quality of life of the user, in average 45% of citizens from the surveyed municipalities have access to basic health insurance services within 500 m. However, the obtained data shows that smaller municipalities have the lowest values. This result confirms the

Table 4 The results of the research (made by authors)

Indicator	Virovitica-Podravina county (1)	Koprivnica-Krizevci county	Varazdin county	Virovitica-Podravina county (2)	Zadarska county	Explanation
Access to basic health insurance (% people)	50%	35%	25%	10%	60%	Up to 50%—poor 50–70%—average Above 70%—good
Encouraging healthy lifestyle (scale from 1 to 5)	1	4	4	4	4	1—poor 2, 3—average 4, 5—good
Traffic accidents (#/1000)	1.25*	20	13	6.68	1*	Above 10—poor 5–10—average Under 5—good
Availability of public transport (% people)	90%	50%	80%	50%	70%	Up to 50%—poor 50–70%—average Above 70%—good
Length of cycling track (% in km)	0%	0%	70%	0.5%	0%	Up to 50%—poor 50–70%—average Above 70%—good
Availability of high-speed Internet (#/100)	0	10	10	33	80	Up to 50—poor 50–70—average Above 70—good
Environmental education (% of schools)	100%	100%	100%	100%	100%	Up to 50%—poor 50–70%—average Above 70%—good
Green area (Ha/1000)	0.15 ha/person*	0.18 ha/person*	0.33 ha/person	0.15 ha/person	0.1 ha/person	Up to 5 m^2 (0.0005 ha)—poor 5–10 m^2 (0.0005–0.001 ha)—average Above 10 m^2 (0.001 ha)—good
Water consumption ((average) no. of litres per	32 lit/person*	30 lit/person	80 lit/person	30 lit/person	120 lit/person	Above 150 lit/person—poor

(continued)

Table 4 (continued)

person per day)						80–150 lit/person—average Under 80 lit/person—good
Population density (#/km²)	40 person/km²	70 person/km²	241 person/km²	63.6 person/km²	52.42 person/km²	Above 200 person/km²—poor 50–200 person/km²—average Under 50 person/km²—good
Municipal waste ((average) weight of waste per capita in tonnes)	0.2 t/person *	0.2 t/person	0.17 t/person	0.2 t/person	0.15 t/person	Above 0.4 t/person—poor 0.2–0.4 t/person—average Under 0.2 t/person—good
Percentage of waste recycling (% in tons)	10.12%	0.00%	40.28%	16.38%	5.00%	Up to 50%—poor 50–70%—average Above 70%—good
Unemployment rate (% people)	7.4%*	25%	8.93%	3.8%	10%	Above 10%—poor 5–10%—average Under 5%—good
Youth unemployment rate (% people)	1.5%*	20%	0%	15%	10%	Above 10%—poor 5–10%—average Under 5%—good
Share of green public procurement (% in HRK)	0%	0%	0%	0 %	0%	Up to 50%—poor 50–70%—average Above 70%—good
Gross domestic product (HRK/person)	3617.00 kn/person	3534.00 kn/person	3800.00 kn/person	3113.00 kn/person	7301.00 kn/person	Under 10,000 kn/person—poor 10,000–50,000 kn/person—average Above 50,000 kn/person—good
Newly registered	1*	0	7	2	3	Under 3—poor 3–5—average Above 5—good

(continued)

Table 4 (continued)

operations (#/1000)						
Entrepreneurial incubators (#/1000)	0	0	0	0.1	1	Under 0.5—poor 0.5–1—average Above 1—good
Tourism intensity (#/1000)	10*	2*	944	100	9	Under 500—poor 500–700—average Above 700—good
Availability of state data (scale from 1 to 5)	3	3	4	4	3	1—poor 2, 3—average 4, 5—good
Smart city policy (scale from 1 to 5)	3	5	2	3	0	1—poor 2, 3—average 4, 5—good
Expenditures by the municipality for a transition towards a smart city (HRK/person)	1.00 kn/person	0.00 kn/person	100.00 kn/person	0.00 kn/person	0.00 kn/person	Under 1000 kn/person—poor 1000–5000 kn/person—average Above 5000 kn/person—good

fact that smaller municipalities do not have access to all health services and in order to use certain services they must reach larger nearby cities (usually county centres), which are often quite far from the municipality itself. Despite the fact that most municipalities do not have access to basic health services, they still estimate that they promote a healthy lifestyle at a high level, with the exception of the municipality in Virovitica-Podravina County, which rates this indicator with the lowest grade-1.

The availability of health services is closely related to the state of traffic in the observed municipalities, but also in the counties. The indicator that shows the number of traffic accidents indicates that more developed areas, such as the northern part of Croatia, have higher number of traffic accidents from less developed areas. Less developed areas have lower number of residents, and this can support the results obtained. Also, when observing the availability of public transport, it is evident that municipalities with smaller population reach up to 90% of the availability of public transport to the population. This can be explained by the fact that they have either bus stations or railway stations that cover all areas of municipalities. The area of the municipality is geographically smaller, which makes it easier to develop public transport and to cover the entire municipality with public transport. When it comes to the percentage of the length of bicycle lanes in kilometres, it can be noticed that three of five municipalities do not have cycling tracks, and that the municipality from the north side of Croatia has 70% street length covered with cycling tracks. This shows the assumption that most of Croatian municipalities don't have a developed network of bicycle paths.

The availability of high-speed internet is ensured in 4 out of 5 municipalities and an average of 27 households out of 100 have this option. In each of the surveyed municipalities, educational institutions maintain education on environmental protection, which is of great importance nowadays given the amount of waste produced. When observing the results of green space per 1000 residents, two municipalities do not keep track of this data. From the municipalities that provided such data, it is evident that the green area per 1000 residents is distributed in accordance with their geographical position. Water consumption per resident varies from municipality to municipality like most data, but the average given by 4 municipalities that have data on water consumption per resident is 65 l/person. Population density varies depending on the number of residents and the geographical area it covers. The most densely populated municipality is the most developed municipality, and that is the municipality from Varazdin County, while the least populated is the municipality from the area of Virovitica-Podravina County. Waste production in each of the observed municipalities is approximately the same, and the average for 4 municipalities is 0.18 t/person. Waste recycling is carried out in four municipalities out of five surveyed, and in all municipalities the recycling percentage is very low, although the municipality from Varazdin County stands out from the rest in waste recycling with its 40.28%. An interesting result is shown by the municipality from Koprivnica-Križevci County with a value of 0%, which is a surprising result because it is the only county (except the City of Zagreb) in which the county centre has a Smart City certificate and where the county develops waste recycling plans. This decrease may indicate the fact that municipal representatives do not know or do not have this information, which may mean a lack of waste management in the area.

The unemployment rate in the surveyed municipalities ranges from 3.8 to 25%, and it can be concluded that more developed municipalities have a lower unemployment rate. The same applies to youth unemployment, where the more developed municipalities have a lower unemployment rate. It is again visible that the Varazdin county municipality has 0% of youth unemployment. Gross domestic product varies, but in this case the highest GDP has the Zadar region municipality (7301.00 km/person), which is not surprising considering that it is a tourist destination, and that tourism takes place throughout the year, especially in the summer months. When it comes to the indicator of newly registered jobs in the previous year, the results are very low, only 11 newly registered jobs at five municipalities in the year 2020. The most of them are in Varazdin county municipality.

All municipalities have a smart city policy. Based on the results obtained by five municipalities, their view is that on average their smart city policy is stimulating on the level of 3.4 on a scale from 1 to 5. For the indicator tourism intensity 2 municipalities have no data on the number of nights per 1000 residents in a year, while in the remaining three municipalities we see quite different numbers, which is again influenced by the position and development of municipalities. The Varazdin county municipality has a high number of overnight stays per year per 1000 residents because it offers various touristic attractions, and has also developed entrepreneurial tourism. A high number of overnight stays is expected in Zadar county municipality, considering that this area is attractive for tourists, especially in the summer months, because it is positioned near the Adriatic Sea. The availability of state data on a scale from one to five received an average score of 3.8, which indicates that work needs to be done on the availability of state data, which municipalities need to operate. Furthermore, when observing the indicator entrepreneurial incubator in 3 municipalities out of 5 there are no entrepreneurial incubators. The municipality in Zadar County owns 1 entrepreneurial incubator per 1000 persons, and the municipality from the area of Virovitica-Podravina County owns only 0.1 per 1000 persons.

The obtained results show that no funds were invested into public procurement with an emphasis on green public procurement in any of the surveyed municipalities, i.e., that no conducted procurement procedure was treated through the elements of green public procurement. Thus, it is evident that municipalities do not have sufficient experience and knowledge in the application of green public procurement. Smart city investments are low, and it is shown by the last answer related to the costs caused by investments in the smart city concept. Very interesting fact is that the Virovitica-Podravina County municipality, which did not invest any funds into smart city concept development, rated their smart city policies with the highest value. 3 of 5 municipalities, made no investments, one municipality invested 100 HRK/person and it is a Varazdin county municipality, in the northern part of Croatia, which again confirms that the municipalities from the north of Croatia are more developed. At the same time, the Virovitica-Podravina county municipality, which belongs to the category of underdeveloped local self-government units, invested 1 HRK/person, which indicates that there is a desire for further progress and development, and the creation of conditions for a better life of locals in a municipality that is less developed.

5 Discussion

Based on the previously shown table, out of a total of 22 indicators, the first municipality located in Virovitica-Podravina County has a total of 7 indicators with good value, 5 indicators with average value and 10 indicators with poor value. Another municipality from the same county has 6 indicators with good value, 5 with average, with 11 with poor value. The Koprivnica-Krizevci County municipality has a total of 5 indicators with good value, 4 with average and 13 with poor value. The Varazdin County municipality shows 10 indicators with good value, 3 with average, and 9 with poor value. Finally, the Zadar County municipality has 8 indicators with good value, 5 with average and 9 with poor value. These results are presented on Fig. 1.

Research conducted on 5 municipalities shows that the municipality that meets the most indicators, but also has quite high results, is a Varazdin county municipality. This is the only municipality in which there are more good values of indicators than poor ones, and only this municipality has the largest number of good values of indicators. Also, this is the one of the two municipalities that participated in the survey, which independently defined the values of the indicators, which indicates the fact that this

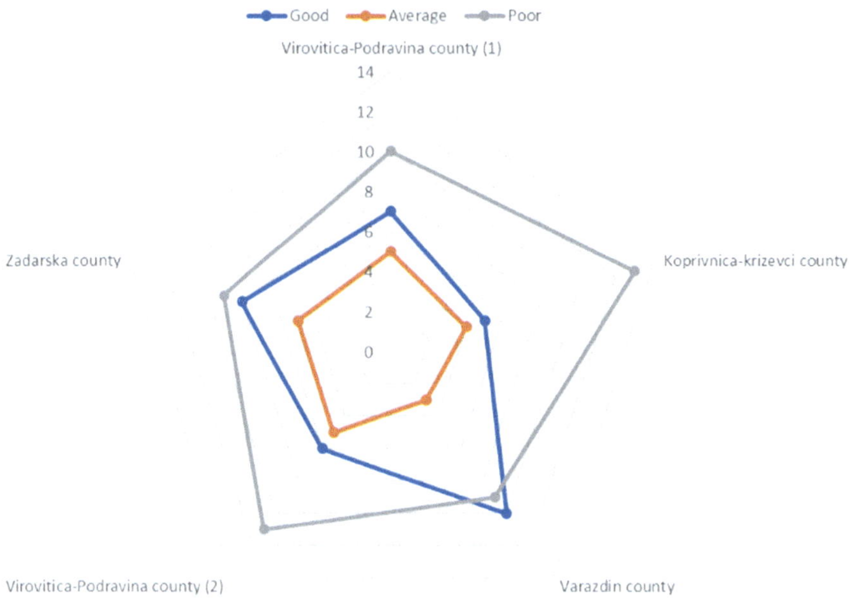

Fig. 1 Number of good, average and poor values by municipality (made by authors)

municipality has all the data. This also shows the fact that within Varazdin County the data is collected and used for decision-making and tracking progress which reflects on the management and development of municipalities within this county.

According to Fig. 1, the worst result is shown by the Koprivnica-Križevci County municipality due to the largest number of poor values of indicators, but very close is the Virovitica-Podravina County municipality (2). Another municipality from Virovitica-Podravina County (1) also shows many poor values of indicators, but there is a significantly higher number of good ones. The Virovitica-Podravina country municipality (1) is also positioned in group 1 according to the index of development of local self-government units, which means that it belongs to the category of the least developed local self-government units in the Republic of Croatia. In the Virovitica-Podravina country municipality (2) is evident that smart city policy is being implemented, but to an insufficient extent the same as in the Koprivnica-Krizevci county municipality. The main problem within the collected data is that there are no results to some of the presented indicators, which indicates that some municipalities do not track data needed to make smart initiatives and projects and that are also needed for making better management choices.

A priority matrix will be used to demonstrate which of the presented municipalities has the best and which the worst result, good, average and poor values need to be weighed through the priority matrix. Therefore, bad value indicators will be weighted with a value of 1, average with a value of 3, and good with a value of 5. Table 5 shows the priority matrix.

Table 5 Priority matrix (made by authors)

Municipalities/counties	Poor value		Average value		Good value		Total
	Weight	1	Weight	3	Weight	5	
Virovitica-Podravina County (1)	10	**10**	5	**15**	7	**35**	60
Virovitica-Podravina County (2)	11	**11**	5	**15**	6	**30**	56
Koprivnica-Krizevci County	13	**13**	4	**12**	5	**25**	50
Varazdin County	9	**9**	3	**9**	10	**50**	68
Zadarska County	9	**9**	5	**15**	8	**40**	64

The priority matrix, shows that the Varazdin County municipality showed the best results, but very close to it is the Zadar County municipality. Varazdin County and the north of Croatia are one of the most developed parts of Croatia, which is why this result is not surprising. Also, Zadar County, as one of the most developed tourist counties, is expected to show high results. The city of Koprivnica as the centre of Koprivnica-Krizevci County is becoming one of the most important smart cities in the Republic of Croatia and is implementing numerous projects in this area, which is a surprising indicator of how the municipality from that county shows the worst results. This could be an indication that investment in smart initiatives is too centralized and that additional investment should be made in the surrounding areas. The results of two municipalities from Virovitica-Podravina County are very interesting. Despite the fact that the result is very similar, and both municipalities show a very similar number of good and poor values of the indicators, there is a significant difference in the availability of data. The Virovitica-Podravina County municipality (2), along with the Varaždin County municipality, has all the values of the required indicators. This indicates that the municipality is focused on development planning and monitoring of implemented activities. Unlike the mentioned municipality, another municipality from the same county provided the least data and most of the values were based on an estimate.

6 Conclusion

The analysis provides an overview of Croatian municipalities which gave data to assess smart city indicators through CITYkeys methodology. 5 municipalities where interviewed and surveyed through this research. Conducted research showed that the most developed municipality by the presented indicators is in Varazdin County. This county is also one of the most developed counties, so it is not surprising that this municipality shows the best results.

The limitations of this research are mainly related to the fact that the values of individual indicators have been estimated and how it is possible to manipulate the given data. It is especially important to emphasize that for some indicators an assessment is needed in assigning values from 1 to 5, where municipalities can be subjective in assessing an individual indicator.

To conclude, the obtained results are not surprizing. In the Republic of Croatia there is an official document in which municipalities and cities are assigned to a certain development group based on the development index. By studying this document in advance, certain conclusions and assumptions were reached. This research can help municipalities move towards the concept of a smart city and serve as a framework for development towards smart municipalities. For further research, more municipalities should be involved in the research and additional correlation between the development of the county in which the municipality is located and the development of municipality should be made.

References

1. Mutavdžija M, Buntak K, Martinčević I (2020) The role of new technologies in smart city management. In: VIII international conference "quality system condition for successful business and competitiveness"
2. Buntak K, Mutavdžija M, Kovačić M (2019) A review on measuring the success of smart city initiatives. Festival kvaliteta Kragujevac
3. Albino V, Berardi U, Dangelico RM (2015) Smart cities: definitions, dimensions, performance, and initiatives. J Urban Technol 22(1):3–21
4. Abdalla W, Renukappa S, Suresh S, Al-Janabi R (2019) Challenges for managing smart cities initiatives: an empirical study. In: 3rd international conference on smart grid and smart cities (ICSGSC)
5. Dameri R (2013) Searching for smart city definition: a comprehensive proposal. Int J Comput Technol
6. Škunca S (2015) Koliko su hrvatski veliki gradovi doista veliki? In: Prostorno planiranje kao čimbenik razvoja u županijama, Zbornik radova
7. Bosch P, Jongeneel S, Rovers V, Neumann H-M, Airaksinen M, Huovila A (2017) CITYkeys indicators for smart city projects and smart cities
8. Huovila A, Airaksinen M, Pinto-Seppä I, Piira K, Bosch P, Penttinen T et al (2017) CITYkeys smart city performance measurement system. Int J Hous Sci Appl 41(2):113–125
9. Airaksinen M, Seppä IP, Huovila A, Neumann HM, Iglar B, Bosch P (2017) Smart city performance measurement framework CITYkeys. In: 2017 International conference on engineering, technology and innovation (ICE/ITMC), June 2017. IEEE, pp 718–723
10. Wendling LA, Huovila A, Castell-Rüdenhausen MZ, Hukkalainen M, Airaksinen M (2018) Benchmarking nature-based solution and smart city assessment schemes against the sustainable development goal indicator framework. Front Environ Sci 6:69
11. Caird SP, Hallett SH (2019) Towards evaluation design for smart city development. J Urban Des 24(2):188–209

Geospatial Analysis of Bicycle Infrastructure in the Function of Traffic Safety: The Case of the City of Zagreb

Dario Oršić, Mario Ćosić, Ljupko Šimunović, and Matija Sikirić

Abstract Traffic infrastructure needs to adapt to ever-increasing demand for bicycle traffic. This will ensure sustainable urban mobility in the future and reduce the negative consequences of modern traffic such as noise, environmental pollution, and harmful gas emissions. Due to the increase in bicycle traffic in Zagreb in recent years, it was necessary to analyze the existing cycling routes, primarily bicycle paths as the most numerous type of bicycle routes in Zagreb. For the analyzed infrastructure, it was necessary to define each segment that can affect the safety of bicycle traffic. In addition, it was important to determine the volume of traffic flow in specific locations where a shared space zone is organized, which may result in more frequent conflicts between bicycle and other traffic flows. The result of the analysis is a geospatial presentation of important segments that affect the safety of bicycle traffic, and their uniformity with the set performance standards, and suggestions for improvement derived from the same.

Keywords Geospatial analysis · Bicycle infrastructure · Traffic safety

1 Introduction

In contrast to the car, the bicycle is an almost ideal solution for accessibility; it does not pollute the environment and enables adequate mobility. In addition, a bicycle takes up much less space compared to a personal car. At least 10–12 bikes fit in one parking space. Thus, encouraging people to increase cycling, at least for some of their daily needs (especially at shorter lengths), will allow greater use of public space for other purposes [1].

In [2] an evaluation of the impact of interventions on bicycle infrastructure in Dutch and Danish cities was conducted. The evaluations were based on before and after condition studies. Research has found that new cycling infrastructure generates new users, and modal change, especially from cars to bicycles, appears to be a more

D. Oršić · M. Ćosić · L. Šimunović · M. Sikirić (✉)
Faculty of Transport and Traffic Sciences, University of Zagreb, Vukelićeva 4, 10000 Zagreb, Croatia
e-mail: msikiric@fpz.unizg.hr

M. Petrović et al. (eds.), *The Science and Development of Transport—ZIRP 2021*,
https://doi.org/10.1007/978-3-030-97528-9_2

marginal effect of bicycle infrastructures. In addition, it was found that cyclists highly appreciate the separated bicycle route designs. The reason for this is the perceived sense of security that such a form of infrastructure offers them. Another important segment, which provides a sense of security, is the reconfiguration of parking areas and locations, i.e., points of conflict with motor traffic.

Similar research was conducted in Kazan, Kaliningrad [3] and Portland [4] where it was also concluded that as much physically segregated cycling infrastructure as possible must be offered to promote cycling. This is particularly pronounced in women who frequently transport children, and a sense of security is a key segment when choosing travel modes [5].

In [6] authors assessed the safety of bicycle traffic with the use of geographic information system "Yandex." A matrix of the level of danger of cyclists was created depending on the volume of traffic, the permitted restriction and the type of intersection. The analysis of the obtained data concluded that in order to achieve a higher level of safety it is necessary to strive for minimizing the risks of collisions with vehicles or pedestrian path in the course of designing the bicycle traffic infrastructure. In doing so, the following recommendations were obtained: maximum spatial separation and disengagement of flows of the cyclists, vehicles and pedestrians; reduction of vehicles' speed at places of crossing or overlapping the bicycle traffic; ensuring sight distance sufficient for reacting and braking at places of approaching a turn, pedestrian crossing, intersection or crossing the other traffic direction and flow; providing continuity of bicycle way and non-stop traffic; use of road pavement with good traction; presence of sufficient illumination of road pavement and signs, especially at potentially dangerous places and in some troublesome zones; and availability of sufficient information at the road (clear lines, signs, placards, signals).

In [7] authors have analyzed cycling injury risks of 14 different bicycle route types and other infrastructure features. The study found that the safest kind of facility, by far, was cycle tracks, which are on-street bicycle lanes that are physically separated from motor vehicles by raised curbs, bollards, or concrete barriers.

In order to determine the current state of the existing cycling infrastructure in the City of Zagreb, a field research was conducted. Based on the collected data, the entire bicycle network was mapped, with all associated segments. In addition, the positive and negative segments of the same were identified, and suggestions for its improvement were made.

The purpose of the research was to define in detail the current state of the cycling network of the City of Zagreb, with the aim of identifying irregularities, possible safety failures for infrastructure users, and identify possible improvements aimed at raising the level of safety of participants. The motivation for conducting the research was the lack of this type of comprehensive analysis in the observed area that the developed database serves as a basis for future research and possible upgrades with new traffic safety indicators.

During the research, the main thesis was set and confirmed:

Comprehensive geospatial analysis and mapping of cycling infrastructure can identify gaps and potential dangers in the network.

The paper is structured as follows. Section 2 presents the scope of the research, defining the key parameters, as well as the current situation at the beginning and end of the research period. Section 3 defines the methodologies of the bicycle infrastructure safety research itself, which include calculating the length of bicycle paths, defining traffic volumes (bicycle and motor traffic) and describing the process of collecting and mapping infrastructure data. Section 4 presents the most important research results, irregularities of the existing cycling infrastructure, as well as examples of good and bad practice found during the field research. This is followed by a discussion of the research results (Sect. 5), and the conclusion and proposal of the next steps in the research extension (Sect. 6).

2 Catchment Area

The area of the City of Zagreb has been determined for conducting the research. City of Zagreb covers a surface area is 641.32 km^2 and is located at 122 ^{m} above sea level. The climate is moderately continental. The total population density in Zagreb is 1259 inhabitants per km^2, while the entire metropolitan area is home to about 807 thousand inhabitants [8].

The area includes the whole area within the Zagreb city limits, which consists of 17 districts, i.e., the area to the farthest points of the city where bicycle paths are located. From Gojko Šušak avenue and Medveščak street in the north to the Zagreb city limit in the south, from the Podsused alley on the west side of the town to Novi Jelkovec and Sesvete on the east side. Since the bicycle paths are the most numerous bicycle routes in Zagreb, the analysis is mostly related to them. There are only two bicycle lanes in Zagreb, while bicycle roads are not yet part of the Zagreb bicycle network. Bicycle trails are not analyzed because they are mostly used for recreation and have no pavement (riverbanks and parks).

Figure 1 shows a map of bicycle paths and bicycle lanes as of May 2021. This analysis also includes all newly built bicycle paths and lanes in Zagreb in 2021. Two-way bicycle paths are displayed in blue, one-way bicycle paths are displayed in red, and the bicycle lanes are marked in green. According to data from the Report on the Cycle Subsystem within the Zagreb City Transport System for the year 2017, the City of Zagreb has 270 km of cycling routes (paths and lanes) [9].

At present, there are several shared space streets where bicycle traffic and motorized traffic are combined. Those include Gajeva and Katančićeva Street (720 m), calmed traffic zone on Držićeva/Vukovarska (1080 m). During 2017, it was carried out in Primorska, Kranjčevićeva, Teslina, Kordunska, Berislavićeva and Kovačićeva Street, which totals 2410 m. By the year 2017, there is 4210 m of shared space roads in Zagreb [9]. In 2018, more shared space zones were created and include Magazinska and Preradovićeva Street and parking lot by the National and University Library. Shared space zones are a good solution for streets with low motor traffic volume where spatial needs could not be fulfilled and the possibility of building a cycle path separated from motor traffic is not possible. Cautious approach should be

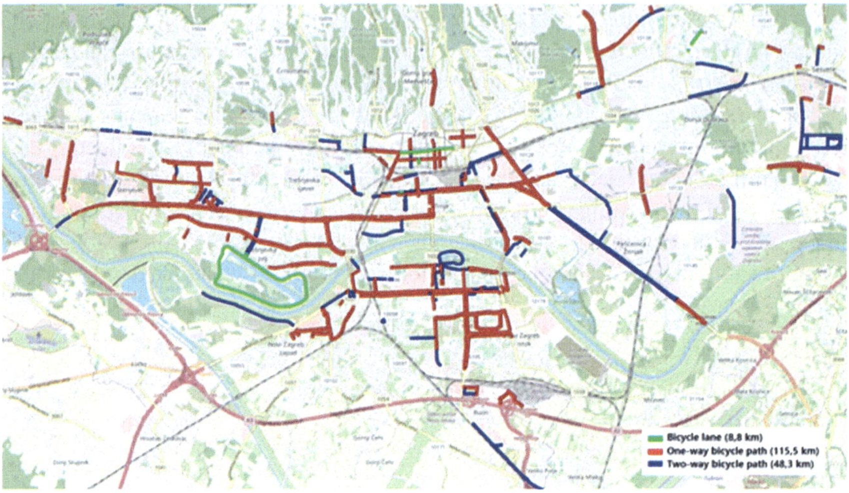

Fig. 1 Bicycle paths and bicycle lanes in Zagreb, May 2021

taken during the introduction of such zones, and before the implementation, motor traffic should be counted at a specific location, otherwise the appropriate conditions for the safe driving of the cyclists will not be met. If the conditions for a safe ride for cyclists are not appropriate, it is questionable how much the introduction of shared spaces will affect the increase in the use of a bicycle as a mean of transport on such streets.

3 Methodology

For the purposes of making an analysis of the safety of cycling infrastructure, the following methods were used:

- Calculation of bicycle path length;
- Determining the traffic volume of bicycle and motor traffic;
- Field research of bicycle infrastructure characteristics;
- Mapping of the bicycle network with all its segments.

3.1 Calculation of Cycling Infrastructure Length

A method has been developed to calculate the length of the existing cycling infrastructure [10]. The method is shown in Fig. 2. The length of the bicycle paths was measured using the Quantum Geographic Information System (QGIS).

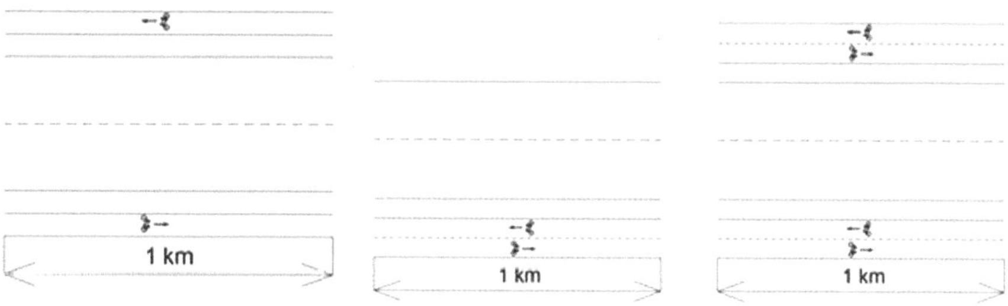

Fig. 2 Methodology used for calculating the length of bicycle paths [10]

If a two-way bicycle path is on either side of the street, then its length was measured for each side separately (e.g., Radnička Road), but if a two-way bicycle path is only on one side of the street then the length was measured only once and not for each direction separately (e.g., Branimirova, Strojarska Street). With official data from the city it is possible that the length of two-way paths on single side of the street between two points was double measured, for each direction separately. Bicycle trails, such as embankments, parks, forest roads and other unpaved surfaces are likely to be included in the total length of cycling routes, but should be in a separate category.

3.2 Volume of Bicycle and Motor Traffic in Shared Space Zones

Traffic counting is the basis for traffic planning and provides an insight into the current state of transport and data that indicate the necessary reconstruction, construction of new traffic directions or other measures to improve existing and future traffic. Traffic counting should be carried out for the purpose of traffic and urban planning, the planning of a prospective traffic network of a larger area or the design of a traffic junction and the eventual reconstruction of the existing transport network and the construction of new traffic routes [4].

Figure 3 shows criteria for selecting cycling routes that depends on the amount of motor traffic and the speed limit. When introducing a new shared space street the following parameters should be considered. At a speed limit of 50 km/h, the number of vehicles must not exceed approximately 375 vehicles per hour, and with a speed limit of 40 km/h 500 vehicles per hour.

Motor traffic counts have been made to check whether the amount of motor traffic corresponds to the mandated selection of cycling route type. The selection of cycling routes depending on the amount of motor traffic and speed is mandated by the Bicycle Infrastructure Regulation [11] and is shown in Fig. 3. Traffic volume is calculated at the intersection of Gajeva Street and Baron Trenko Street, Magazinska and Kranjčevićeva Street. It was also checked whether the drivers respect the "bike box" and whether they stop at a proper distance without entering with their vehicle

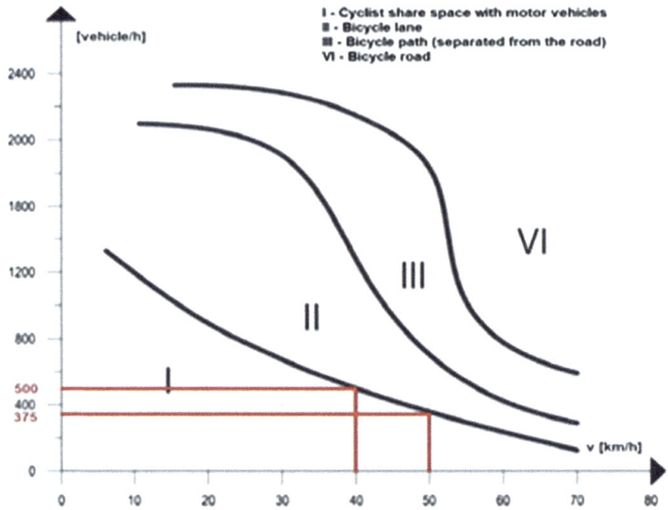

Fig. 3 Criteria for choosing a cycling route type [11]

inside "bike box." It was noticed that 5.5% of the drivers stop their vehicles inside "bike box" at location 1 and 3.2% of the drivers at location 2. In real-life conditions, it turned out that "bike boxes" are not functional unless there is a bicycle lane that leads to "bike box" and allows cyclists a clear and secure path to the box.

Traffic volume was measured in a one-week period throughout the 24 h. At Location 1: Magazinska Street, counting started on 10/30/2018 and ended on 11/07/2018 and oncoming traffic volume was measured (one-way street). At Location 2: Kranjčevićeva Street, counting started on 11/08/2018 and ended on 11/14/2018. Cross-sectional traffic volume was measured (two-way street). 60 min was taken as a time interval.

3.3 Field Research and Mapping of Bicycle Infrastructure Segments

Bicycle paths were mapped by conducting a field research on the bicycle network. Field research was conducted in the period November 2018 and May 2021, in order to obtain a comparison of the situation and progress. Field research was conducted in November 2018 and May 2021, to obtain a comparison of the situation and progress. In addition to the mapping itself, network segments (Table 1) determined by the Bicycle Infrastructure Regulation [11] were recorded.

Since the pedestrian crossings in the City of Zagreb are everywhere marked in an unique way (using only "continental" type marking), bicycle crossings should also look the same in terms of design. They should be marked with red paint with no

Table 1 Types of bicycle network irregularities [11]

Type of irregularity	
Unmarked/irregularly marked bicycle crossings	Irregularly placed manhole covers
Insufficient safe zones between bicycle path profile and object/from the curb/from parallel-parked vehicles	Transition zones between bicycle path and motor traffic/absence of bike ramps at underpasses
Unmarked parts of bicycle path/bicycle path breaks	Marking irregularities
Insufficient width/height of the bicycle path profile	Traffic lights not paired with bicycle crossings
Damaged or uneven driving surface/high curbs	Illegal parking on bicycle paths
Irregular routing through the bus stop zone	

exception in order to make the drivers of the motor vehicles informed unequivocally and clearly when approaching bicycle crossing.

4 Analysis Results

4.1 Bicycle Path Length

The total length of bicycle paths and lanes of 162.6 km was determined in the field, of which 111.8 km of one-way bicycle paths and 49.4 km of two-way bicycle paths and 1.4 km of bicycle lanes. Data refers to November 2018 and includes an all-new bicycle path on Radnička Road. The deviations of this data from the official data from the City of Zagreb can be explained by a different way of length calculation, as defined in more detail in the previous chapter of the methodology.

4.2 Shared Space Zones

Figures 4 and 5 show traffic flow data based on the traffic volume and time passed. Volume is displayed in vehicles per hour and time period of one week was taken. Drops in volume are visible during the weekends (Saturday, November 3, and Sunday, November 4) and holidays (November 1) and to be expected. On both locations during the other days the traffic volume is above mandated mark for shared space streets as shown on Fig. 3. On 2 November there is a slight drop in volume because it is the day in-between the holiday and the weekend.

Figure 6 shows the measured vehicle speeds in Kranjčevićeva and Magazinska Street. Kranjčevićeva Street is a two-lane two-way street so the cross-sectional volume and speed were measured (location 2 on Fig. 4). The display shows that

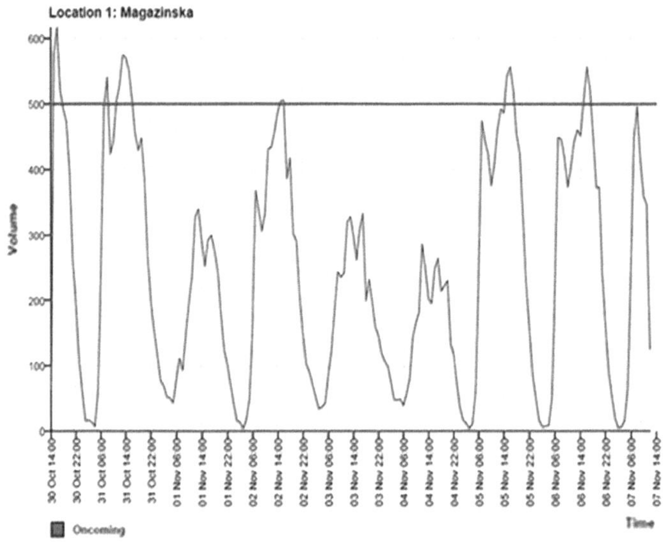

Fig. 4 Traffic flow data, location Magazinska

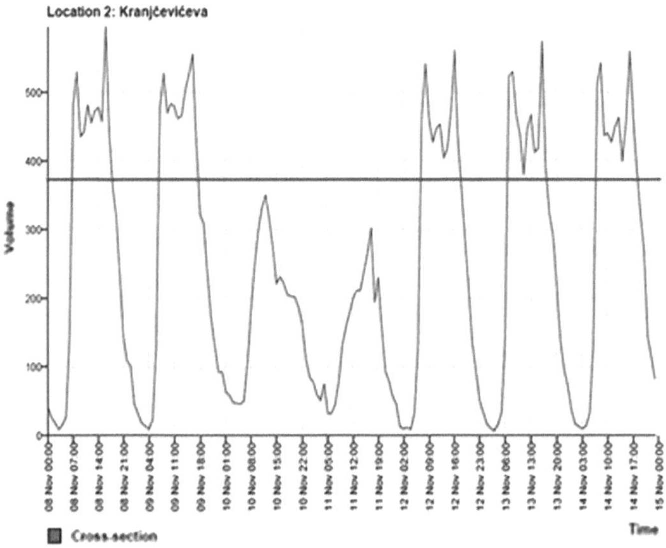

Fig. 5 Traffic flow data, location Kranjčevićeva

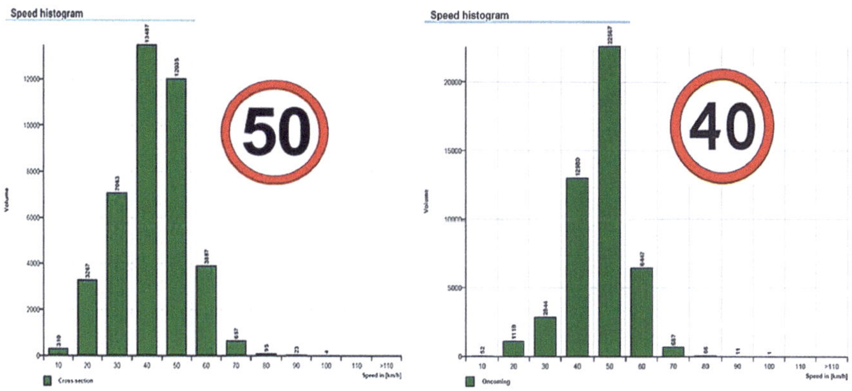

a) Kranjčevićeva street

a) Magazinska street

Fig. 6 Motor traffic volume and vehicle speeds

88.57% of drivers adhere to the speed limit on the observed road (recorded vehicle speeds: <20 km/h = 8.76%; 20–30 km/h = 17.25%; 30–40 km/h = 33.06%; 40–50 km/h = 29.49%). However, the problem for traffic safety is represented by 11.43% of drivers who recorded movement above the permitted limit. This is particularly worrying for 0.3% of drivers with speeds ≥80 km/h. Although these drivers represent a small percentage of the total traffic flow, they are a group of drivers whose careless driving can lead to collisions with severe consequences. Magazinska is a one-way street with single shared lane and one lane reserved for parallel parking (location 1 on Fig. 4).

On the observed road, the speed limit was set at 40 km/h. The analysis shows that only 36.3% of drivers adhere to the speed limit (recorded vehicle speeds: <20 km/h = 2.5%; 20–30 km/h = 6.08%; 30–40 km/h = 27.75%). Most drivers, i.e., 48.25% of them, drive slightly above the limit (≤50 km/h), which, although often tolerated by law enforcement officers, represents an additional risk for vulnerable road users. A special danger is posed by 15.42% of drivers for whom speeds over 50 km/h were recorded, especially 0.17% for whom speeds ≥80 km/h were recorded during the observation. Cyclists are left in an unpleasant position between speeding vehicles passing from the one side and parked vehicles on the other side.

4.3 Bicycle Network Irregularities

There are many irregularities found on the bicycle network and they are grouped into 11 categories. Bicycle paths are analyzed by cycling on the bicycle network. Some of the more significant irregularities are explained and displayed in pictures. Irregularities found on Zagreb bicycle network with total count for each category are shown in Table 2 and more significant ones are shown in Fig. 7.

Table 2 Recorded irregularities on the cycling infrastructure in Zagreb, 2018 and 2021

Type of irregularity	State 2018	State 2021
Unmarked/irregularly marked bicycle crossings	650/601 crossing/s	573/601 crossings
Insufficient safe zones between bicycle path profile and object/from the curb/from parallel-parked vehicles	526 objects/81 location/17 locations	516 objects/n.a./17 locations
Unmarked parts of bicycle path/bicycle path breaks	14.7 km/8.5 km	n.a.
Insufficient width/height of the bicycle path profile	16.2 km/21 location	n.a.
Damaged or uneven driving surface/high curbs	154 locations/250 curbs	n.a.
Irregular routing through the bus stop zone	72 bus stops	67 bus stops
Irregularly placed manhole covers	54 covers	54 covers
Transition zones between bicycle path and motor traffic/absence of bike ramps at underpasses	None/8 underpasses	n.a.
Marking irregularities	//	//
Traffic lights not paired with bicycle crossings	152 crossings at 51 intersection	n.a.
Illegal parking on bicycle paths	19 locations	19 locations

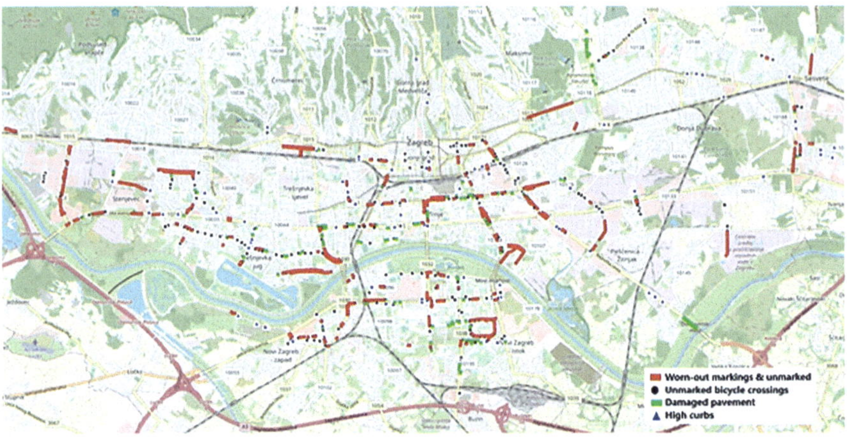

Fig. 7 Display of recorded irregularities of the bicycle network, 2021

At present, there are 573 unmarked crossings in Zagreb, where potentially dangerous places are included, which are not crossings but must be marked with red paint (entrance to parking lots, side entrances etc.). Bicycle paths are often broken on intersections and cyclists must past them as pedestrians by pushing the bicycle across the intersection (Fig. 7).

In addition, traffic lights with bicycle symbols are placed on multiple intersections even if there are no bicycle crossings marked. For those reasons, cyclists often do not take off from the bicycle when crossing the street, which may lead to an accident.

Maybe even bigger problem represent irregularly marked crossings. Some of them are marked using red paint and some without red paint. Some of them even though are marked with red paint do not have proper design and it is questionable if they are considered legal from the police aspect. There are six different ways of bicycle crossing markings spotted on the bicycle network, which may leave both cyclists and motor vehicle drivers in doubt how to act when approaching one. It also may lead to difficulties when determining culprit if accident occurs. All of the above leaves both cyclists and motor vehicle drivers in a "gray zone" and depends on the individual opinion of the police.

Insufficient safe zones correspond to objects or obstacles placed too close to the bicycle path and often directly on the bicycle path (light poles, traffic lights, fire hydrants, bus shelters etc.) Safe zone from the curb should be minimum 0.5 m. At some locations, bicycle path is too close to the traffic passing directly by the cyclists such as Most slobode where buses and trucks may even cause side slipstream and make cyclist to fall off the bicycle. Routing bicycle path to close to parallel-parked vehicles may also result in an accident when opening car doors (dooring).

Figure 8 shows unmarked bicycle crossings at several high-frequency bicycle corridors. At the intersection of Selska and Zagrebačka Avenue there is also a problem of lack of space where pedestrians and cyclists can wait for green light during the rush hours, since the phase on the traffic light allows the transition to only half of the

a) Selska/Zagrebačka Avenue b) Vukovarska/Strojarska

Fig. 8 Unmarked bicycle crossings at the high-frequency bicycle corridor intersections

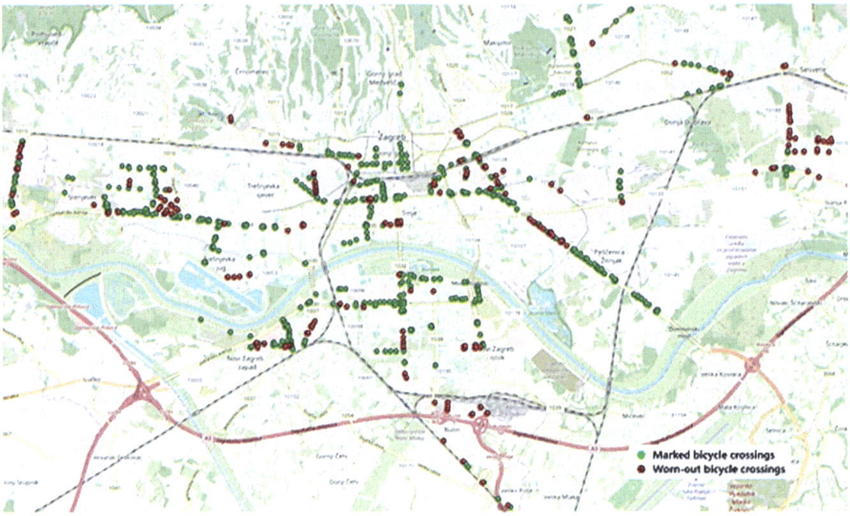

Fig. 9 Overview of the state of bicycle crossings, 2021

intersection. The space could expand if a bicycle crossing is marked by the pedestrian crossing (Fig. 9).

When analyzing bicycle paths, the integrity of the path has been taken as an important factor, including crossings at intersections that should connect these paths. In all locations (intersections and pedestrian crossings) where there is a bicycle path before and after the intersection and the bicycle paths opposite to one or the other, a bicycle crossing that will allow for a continuous bicycle ride should also be marked. All locations where this is not the case are marked with an unmarked crossing label. This is an important safety element and is defined in Article 6 of the Bicycle Infrastructure Regulation: "The points of conflict between motor traffic and cyclists that cannot be avoided (at intersections and crossings) should be marked by traffic signaling so that all drivers, not just cyclists, are aware of the risk and can adjust their behavior" [11].

In addition to breaks, there are unmarked parts of the bicycle paths in the total length of 14.7 km. Those are parts where the markings are worn-out and have not been refurbished or locations where no new horizontal signalization is marked after the construction works were completed (Fig. 11). However, in 2020 and 2021 multiple crossings were marked. As an example, Vukovarska Street was taken as shown on Fig. 10. Poor marking often leads to conflict between pedestrians and cyclists and should be taken seriously. Cyclist can achieve speeds, which are multiple higher than pedestrian speeds, especially with electric bicycles, which are becoming more popular so proper, and clearly visible markings are necessary.

To avoid conflict on bus stops between pedestrians and cyclists, it is planned to route a bicycle path behind a bus shelter or directly over the bus lay-by. There were 72 locations in Zagreb in 2018 and 67 in 2021 where the routing of the bicycle path

Fig. 10 State on Vukovarska Street 2018 versus 2021

a) Unmarked bicycle path (worn -out markings)

b) Unmarked bicycle path (due to constructions)

Fig. 11 Unmarked bicycle paths

in the bus zone was performed incorrectly. There are also several locations where bus shelters are placed directly on the bicycle path. On some parts of the bicycle network corrections were made, like on Branimirova Street (Fig. 12). As a good example, the Radnička Road can be mentioned where the routing of bicycle paths in the bus stop zones was taken into account when planning the route and performed in a proper manner.

In order to make the corrections systematically, the categorization of the bicycle paths was made in four categories. Since the Bicycle Infrastructure Regulation is rather strict, a small percentage of paths are fully in line with regulation. Despite this, a large number of paths can be redesigned to meet regulation with minimal corrections. It has been established that there are 13.7 km of paths in category 4 (not feasible in current conditions), 14 km of paths in category 3 (which can be corrected with greater corrections) and 111.2 km of paths in category 2 (minimum corrections). Total amount of regular bicycle paths is 19.4 km, which is about 12% of the total length of all bicycle paths in Zagreb (Fig. 13).

Fig. 12 State on Branimirova Street 2018 versus 2021

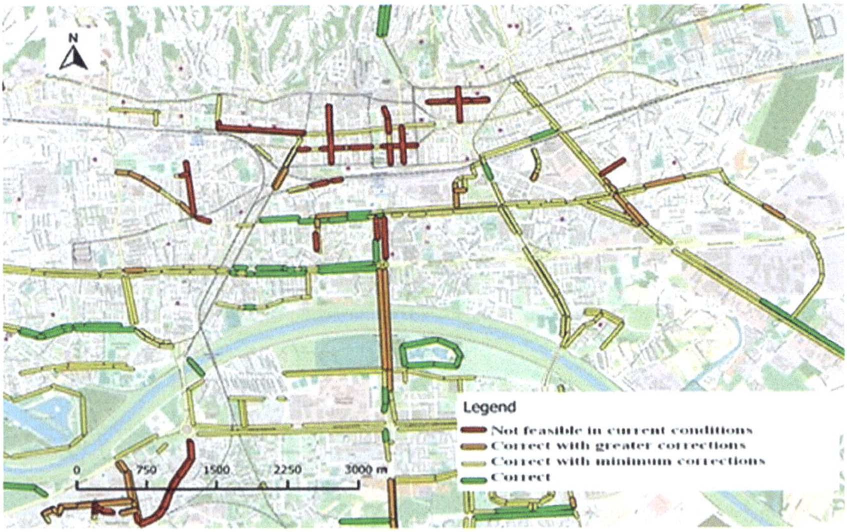

Fig. 13 Categorization of cycling paths and lanes according to irregularities (July 2018) [10]

5 Discussion

This analysis has proven that Zagreb bicycle network has multiple disadvantages, some of serious nature which may endanger cyclists and must be corrected in shortest possible time period like marking bicycle crossings at some key intersections. Bicycle paths are not properly integrated and connected. Quality of the paths may vary between wide paths with high quality pavement, which provide safe and comfortable ride to narrow, bumpy and unmarked paths. Quality paths represent only small segment of the bicycle network and those paths are often not connected with the rest of bicycle network.

General impression could be that bicycle paths are not improved or built with priorities in mind. Thus, there are paths in the city center with high volume of cyclists per hour, which are narrow, to close to parked vehicles with unmarked bicycle crossings at busy intersections and on the other side, there are newly built paths within city suburbs, which are less used and needed. Most of the improvements on the paths include only painting (marking) and rarely any construction work such as removing light poles or other objects in the way of bicycle path. Pedestrian crossings are being refurbished without adding bicycle crossings at the locations where bicycle crossings are needed.

There is often no consistency regarding direction of travel, e.g., on Vukovarska which is one of the main corridors for cyclists, there are segments where one-way path becomes two-way path and then suddenly stops without any traffic sign or transition zone that would allow cyclist to merge between bicycle path and motor traffic. Such infrastructure encourages breaking the law, endangers cyclists and pedestrians and reduces overall traffic safety. Much can be done with small improvements like lowering curbs at the intersections or installing ramps at underpasses and thus raising the comfort and safety of the cyclists with minimal investments. The lack of space in urban areas was tried to overcome by introducing shared space zones and by painting "sharrows" on the streets but without considering the amount of traffic passing on such streets. This analysis should provide data, which is crucial for improving bicycle infrastructure in Zagreb and indicate potential "weak spots" in the infrastructure where safety measures are required. All of the above-mentioned improvements may contribute to bicycle share increase in modal split and partially solve or reduce significant problems regarding parking spaces, pollution, noise and traffic jams.

The methodology used can be applied in almost every city and area of observation. The application of the methodology depends on the legislative framework of a particular environment, on which the methodology for defining irregularities on the network is based, and it can be adjusted according to its regulations. Accordingly, the results of the research can be compared with environments that share similar parameters, and continuous monitoring to determine the impact of changes in the network on the level of traffic safety, and growth/decline in the number of users of the same. This is especially important as a segment in recognizing good practice, which can then be mirrored to environments with similar characteristics.

6 Conclusion

Although only a few bicycle paths in Zagreb are in line with regulation, large amount of paths can be altered with no major interventions. A minimum of at least 100 km paths may significantly be improved if markings were to be refurbished and damaged surfaces to be repaved or repaired. Bicycle crossings should be marked on key intersections with high volume of cyclists per hour. Irregularly marked crossings may also be easily redesigned to meet regulations in most cases. Larger interventions like removing light poles which require higher financial resources should be considered on the locations where most of the path on route is in line with regulation. Bicycle paths with too many irregularities where certain requirements cannot be met should be erased and other solutions should be taken in reconsideration (shared space, traffic reorganization, parking spaces removal). It is necessary to constantly educate and warn all traffic participants when it comes to bicycle traffic. It is also needed to determine the direction of parking policies in city centers and solve delivery parking space issues. More attention should be given to safety of the cyclists when introducing shared space zones on streets, building new or redesigning old bicycle paths. All financial investments invested in improving the safety of cyclists are a step closer to zero accident vision and will pay off in the future through lower external costs.

The next step in the research would be to make a detailed analysis of the locations of traffic accidents in which cyclists participated, and to overlap it with the obtained data from the analysis. In addition, it would be necessary to compare the frequency of accidents at the analyzed sites in 2018, which were marked as dangerous and in the meantime removed, and thus determine the success of the applied changes to the infrastructure.

References

1. Šimunović L, Ćosić M (2015) Nemotorizirani promet. Faculty of Transport and Traffic Sciences, Zagreb
2. Van Goeverden K, Nielsen TS, Harder H, van Nes R (2015) Interventions in bicycle infrastructure, lessons from Dutch and Danish cases. In: 18th Euro working group on transportation. Transp Res Procedia 10:403–412. Elsevier, Netherlands
3. Trofimenko Y, Shashina E (2017) Methodology and results of assessing safety of bicycle infrastructure in Russian cities. In: 12th international conference "organization and traffic safety management in large cities". Transp Res Procedia 20:653–658
4. Monsere CM, McNeil N, Dill J (2012) Multiuser perspectives on separated, on-street bicycle infrastructure. Transp Res Rec J Transp Res Board 2314(1):22–30
5. Reynolds CCO, Harris MA, Teschke K, Cripton PA, Winters M (2009) The impact of transportation infrastructure on bicycling injuries and crashes: a review of the literature. Environ Health 8(47):1–19
6. Garrard J, Rose G, Lo SK (2008) Promoting transportation cycling for women: the role of bicycle infrastructure. Prev Med 46(1):55–59
7. Teschke K, Harris MA, Raynolds CCO, Winters M, Babul S, Chipman M, Cusimano MD, Brubacher JR, Hunte G, Friedman SM, Monro M, Shen H, Vernich L, Cripton PA (2012) Route infrastructure and the risk of injuries to bicyclists: a case-crossover study. Am J Public Health 102(12):2336–2343
8. City of Zagreb (2020) City office for strategic planning and city development: Statistički ljetopis Grada Zagreba 2020. City of Zagreb, Zagreb
9. City of Zagreb (2017) Izvješće o biciklističkom podsustavu unutar prometnog sustava Grada Zagreba 2017. http://web.zagreb.hr/sjednice/2017/sjednice_skupstine_2017.nsf/0/DB6B4E6C5200A4BFC125822E0040F9A2/$FILE/02%20Izvje%C5%A1%C4%87e.PDF. Accessed 10 Apr 2021

10. Oršić D (2018) Geospatial analysis of cycling routes in the City of Zagreb as a function of safety. Master's thesis, University of Zagreb, Faculty of Transport and Traffic Sciences
11. Republic of Croatia, Ministry of Maritime Affairs, Transport and Infrastructure (2016) Pravilnik o biciklističkoj infrastrukturi NN 28/2016. Narodne novine. https://narodne-novine.nn.hr/clanci/sluzbeni/dodatni/439893.pdf. Accessed 4 Apr 2021

University Infrastructure, Sustainable, and Equitable Accessibility? Study Case: Manizales–Villamaría (Colombia) Metropolitan Area

Juan M. Gómez, Carlos A. Moncada, and Diego A. Escobar

Abstract This research evaluates the conditions of geographical accessibility to the educational infrastructure of Manizales and Villamaría (Colombia) metropolitan area, comparing whether private vehicle or public transport is used, allowing to establish the existence of inequality in access to this urban service. The study is developed using methodologies of integral average accessibility, taking the average travel time needed to reach such infrastructure as a comparative measure. The geostatistical model Kriging Ordinary with linear semivariogram is implemented, facilitating the interpretation by allowing its presentation in a graphical form. Additionally, an analysis of population coverage and socioeconomic equity is performed comparing the coverage between different socioeconomic strata. The main results show (1) the public transport system is at a clear disadvantage compared to private transport, given the spatial layout of the infrastructure; (2) the higher socioeconomic strata have better levels of coverage of this urban service; and (3) higher education has poor levels of accessibility.

Keywords Accessibility · Colombia · Coverage · Inequity · Sustainability · Transport

J. M. Gómez (✉) · D. A. Escobar
Facultad de Ingeniería y Arquitectura, Departamento de Ingeniería Civil, Grupo de Investigación en Movilidad Sostenible, Universidad Nacional de Colombia, Sede Manizales, Campus la Nubia, Carrera 37—Calle 94, Bloque S2—oficina 208, 170003 Manizales, Colombia
e-mail: jumgomezmo@unal.edu.co

D. A. Escobar
e-mail: daescobarga@unal.edu.co

C. A. Moncada
Facultad de Ingeniería, Departamento de Ingeniería Civil y Agrícola, Programa de Investigación en Tránsito y Transporte—PIT, Universidad Nacional de Colombia, Sede Bogotá, Ciudad Universitaria edificio 214 oficina 417, 111321 Bogotá, Colombia
e-mail: camoncadaa@unal.edu.co

M. Petrović et al. (eds.), *The Science and Development of Transport—ZIRP 2021*,
https://doi.org/10.1007/978-3-030-97528-9_3

1 Introduction

Due to the transformations and economic changes in recent years, cities have grown and the spatial distribution of the population has changed, resulting in new populated sectors in the expansion areas. However, educational facilities at different levels, for the most part, remain the same as in past decades, thus having sectors with more demand for education but maintaining a low supply of this urban service. This means that in these areas with a growing population, people are forced to seek educational services in remote areas of the city, for which they have to make longer and more time-consuming journeys. Education is a fundamental right and should be available to the entire population regardless of their socioeconomic status (stratum 1 = low economic capacity, stratum 6 = high economic capacity). Manizales is Caldas Department's capital and together with Villamaría comprises the metropolitan conurbation area, located on the central mountain range (Fig. 1) at an altitude of 2150 m.a.s.l., where 513,589 people live [1] in an area of 571.84 km^2 [2]. The metropolitan area is developed in a rather abrupt topography, a situation that leads to serious difficulties in the process of expansion and urbanization.

Hansen [3] defines the term accessibility: "The potential opportunities for interaction"; from this, multiple approaches to the subject have been developed, expanding the term, and providing new points of view and possibilities of application. Geographic accessibility allows, in general terms, to establish the ease of communication between different points of interest [4] with a mode of transport [5], which adds to mobility as an influential variable in the individual's capacity to make

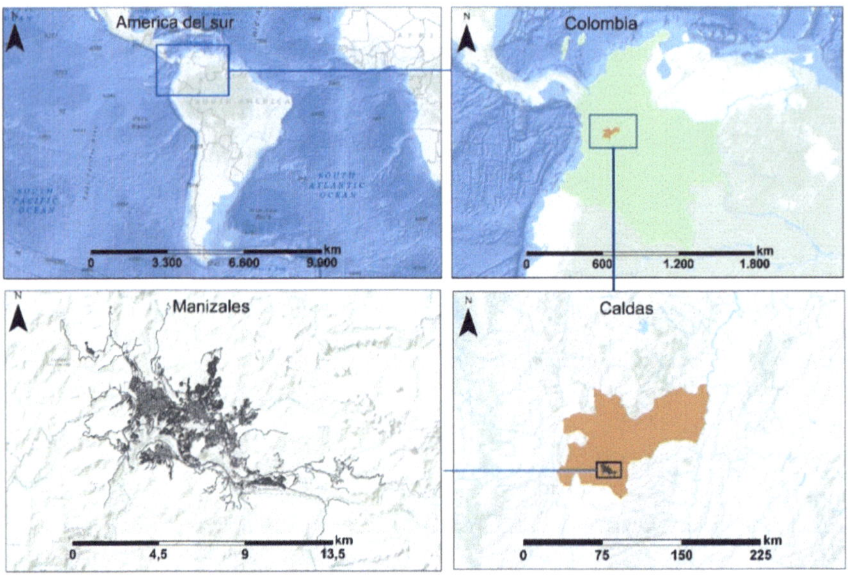

Fig. 1 Manizales city location

daily trips [4]. Accessibility analyses are becoming increasingly complex, seeking the optimal combination of land use, transport modes, and development [6], and methodologies based on regional infrastructure availability [7], gravity model [8], and floating catchment area models [9], etc., have been developed. Due to the great importance of educational services for all types of population and its designation as a basic need by the United Nations [10], accessibility studies on the location and coverage of educational centers have been carried out in different parts of the world to determine the main characteristics that influence access to this urban service. For example, in Argentina, the accessibility assessment was based on linear distance to schools in Necochea and Quequén [11]; in Ireland, they have relied on distance measures for their accessibility calculations [12]; in Canada, taking variables in addition to distance such as parental schooling and economic factors [13]; in the UK, evaluating the expansion of universities and its impact on different economic levels of the population [14]; in England, using the distance between homes of origin and universities [15]; in Portugal, analyzing different variables involved in decisions to study outside the place of origin [16]; in the Netherlands, analyzing the impact of accessibility to universities and the influence of school background [17]; in Germany, including socioeconomic variables [18]; in Colombia including education as one of the primary activities of a population [19], and involving accessibility in equity assessments according to the spatial distribution of educational service [20, 21].

Table 1 presents a bibliographic review of the topics covered in this article, organizing in chronological order research that in some way contribute knowledge for the realization of this research. It is evident how the measurement of distance through a network (Net) is increasingly used over the years over the Euclidean (Eu) distance; in addition, the travel time (TV) is little used despite presenting the most realistic measurements. Different types of accessibility have been used to evaluate accessibility to facilities within cities, such as relative accessibility (Rl), potential accessibility (Po), integral accessibility (In), opportunity accessibility (Op), geographic accessibility (Gr), and the two-step floating catchment area method (2SFCA). This has been done mainly to measure the access conditions in a certain mode of transport (Pr: Private; Pub: Public; Wa: Walking), few cases have compared the travel time values presented between two modes of transport to the same destination, and no case so far presents the ratio between modes as the one presented in this research, filling a gap in the methodology related to the issue of geographic accessibility.

2 Methodology

The methodology used consist of seven phases, which are described below.

Table 1 Literature review

Year	Author	Social equity	Education centers	Measuring	Acc. type	Transport mode	Mode ratio
1959	Hansen (1959)				Rl, Po	Pr	
1998	Talen y Anselin (1998)			Net	Po		
2004	Smoyer-Tomic (2004)	x		Eu	In		
2004	Blanden (2004)	x	X				
2005	Tsou (2005)		X	Net	In		
2005	Shen (2005)	x		TV	Op	Pr, Pub	
2006	Omer (2006)	x		Eu	In		
2006	Frenette (2006)	x	X	Eu			
2006	Sá (2006)	x	X	Eu	Gr		
2009	Gómez (2009)	x	X				
2010	Villanueba (2010)		X	Eu	In	Pub	
2010	Spiess (2010)	x	X	Eu	In		
2011	Chang (2011)	x		Net	Op	Pr, Wa	
2012	Delmelle (2012)	x		Net		Pub, Wa	
2012	Avendaño (2012)		x	Net	Po		
2012	Gibbons (2012)	x	X	Net	In		
2013	Cullinan (2013)	x	X	Net	In		
2013	Mao (2013)	x		TV	2SFCA	Pr, Pub	
2013	Wang (2013)			Net, TV	Po	Pr	
2014	Taleai (2014)		X	Net	Op		
2015	Walsh (2015)		X	Net	Gr		
2015	Rahman (2015)	x		Eu			
2016	Gao (2016)	x	X	Net			
2016	Dadashpor (2016)	x		Eu			
2016	Lucas (2016)	x			Gr		
2016	Younes (2016)	x	X	Net	In	Pr	
2017	Escobar (2017)	x	X	Net	In	Pr	
2017	Shen (2017)	x		Net	2SFCA	Wa	
2018	Almohamad (2018)	x		Net	Op		
2018	Montoya (2018)	x	X	Net	In	Pr	
2019	Zhang (2019)			TV	In	Pr	

2.1 Collection of Supporting Information

Three main databases were used for the research development (i) sociodemographic and socioeconomic information of Manizales–Villamaría metropolitan area, which was obtained from official reports of Manizales Administration's Office; (ii) database of higher education institutions (24 universities), based on micro-data published by the National Department of Statistics (Departamento Nacional de Estadística—DANE); (iii) transport infrastructure network, both private and public.

2.2 Updating, Adjustment, and Calibration of the Transport Infrastructure Network

The investment in infrastructure has not stopped in recent years, which requires an update of the digitized transport infrastructure network with the most recent works for an adequate calculation of travel times and accessibility levels. The transport infrastructure network was updated, adjusted, and calibrated from the supporting database to which operational speeds were uploaded from GPS data for both private [21, 22] and public transport [23]. The network is composed of more than 12 thousand arches and 8 thousand nodes. The network has a length of 685 km, main roads (10%), secondary roads (9%), collector (14%), and local (53%).

2.3 Georeferencing Universities

The existence of the universities registered in the DANE database is verified through fieldwork. The area occupied by each university was digitized and the data necessary for the research was uploaded. A total of 13 universities offering vocational education and 11 universities offering technical education were found.

2.4 Vector Calculation of Travel Times

From the travel times between each pair of nodes in minutes, the minimum travel times in the transport infrastructure network are calculated, creating a matrix ($n \times 1$), for each case, private mode and public mode. Travel times between different start nodes (i) and arrival nodes (j) are calculated using the Dijkstra algorithm, better known as the minimum paths algorithm.

2.5 Geostatistical Computation and Isochronous Curves

The types of geographical accessibility to be used in this analysis are global average and integral average accessibility. The first type allows evaluating the general accessibility conditions provided by the infrastructure network in the city, identifying the isochronous curve on which each university is located; such analysis is done for public and private modes of transport. The second type allows evaluating the accessibility conditions that the infrastructure network provides to reach the particular university points, which will be the basis to identify the sectors of the city in which it is more efficient to reach a university institution using public or private mode.

Once the set is built, the labels *P* are constructed and assigned the values $P_o = 0$ and $P_i = \alpha$ if *i* is different from 0, and the variable $t_i = 0$ is also generated. Then, the last vertex incorporated to the set *S* is established as (*y*), in such a way that for every vertex *z* not included in *S*, $P_z = \min\{P_z, P_y + d_{yz}\}$ is calculated; if $P_y + d_{yz} < P_z$, $t_z = y$, then z^* is determined from among all the *z* vertices that comply with $P_z^* = \min\{P_z\}$. If $P_z = \alpha$ the search process is finished, and it is assumed that there is no path of finite length between *O* and the vertices of the set *S*. Otherwise, if there is a connection, the process is continued, and z^* is incorporated into the set *S*, where P_z^* will be the value of the minimum distance between *z* and *O*. Finally, if $z = z^*$ and $F = z^*$, the iteration process is finished; otherwise, the cycle continues. Subsequently, it is necessary to expand the point calculations to geographic areas by applying geostatistical models. This is done through the Geostatical Wizard tool of the ArcMap program, which, through the Ordinary Kriging method (Eq. 1) with linear semivariogram, allows estimating or extending values to areas without information considering that the distance between the defined points shows a spatial correlation that allows explaining the behavior of a surface [24].

$$Z(S_0) = \sum_{i=1}^{N} \lambda i Z(S_i) \tag{1}$$

With $Z(S_i)$—measured value at the location *i*; λ—weighting for the mean value at the location *i*; S_0—prediction location and *N*—number of measured values. Once the correlation process is finished, the tool displays the different levels of accessibility in the study area in the form of isochronous curves.

2.6 Coverage Analysis of Isochronous Curves

The isochronous curves are exported as a polygon-type layer, with information on travel times loaded in its attribute table. These layers allow making intersections with the layer containing the sociodemographic information, thus obtaining the basis for calculating the percentage coverage ogives.

2.7 A Comparison of Accessibility Conditions Between Public and Private Mode

In this phase, two comparisons are analyzed: the first is achieved by comparing the spatial distribution of the universities, according to the range of isochronous curves, both for private and public modes; the second is the comparison of the results based on the average comprehensive accessibility, finding which sectors of the city refer to better accessibility conditions in public mode. This analysis is the basis for establishing whether there is greater or lesser inequality in the conditions of access to the university service in the city.

3 Results and Discussion

3.1 Overall Average Accessibility

The results obtained about the overall average accessibility conditions show the average-travel-time curves in which the universities are located and in which time ranges they are located the most, both for public and private modes. The results corresponding to the public transport mode are presented in Fig. 2, where the travel time ranges reach the red shades in times of up to 105 min, a considerably higher time than the upper limit resulting from the analysis of private transport. However, in the areas with a higher degree of urbanization, the green colors prevail, corresponding to times of less than 30 min, leaving the peripheral sectors with the highest values. The area with the lowest travel times is observed in the city center, extending along the main roads, especially with Avenida Kevin Angel, located to the northwest.

The resulting times for private transport modes range from 15 to 55 min and are shown in Fig. 3. Much of the study area is under 22 min, with the lowest times in the central area. The influence of the main roads on accessibility levels is evident, causing the sectors located in their vicinity to have lower average travel times. Reduced levels of accessibility are also demonstrated in the municipality of Villamaría due to the existence of a single point of connection for the entry and exit of its inhabitants.

When crossing the location of the universities with the global average accessibility curves for each mode of transport, it is observed that these are located on average-travel-time curves of between 17.5 and 40 min. Likewise, when comparing accessibility conditions between public and private transport, 8% of the universities report better access conditions for public transport, 38% report equal access conditions for both modes of transport, and 54% report better access conditions for private transport, which is precisely the mode of transport that is considered the less sustainable and the one that limited number of students are likely to use.

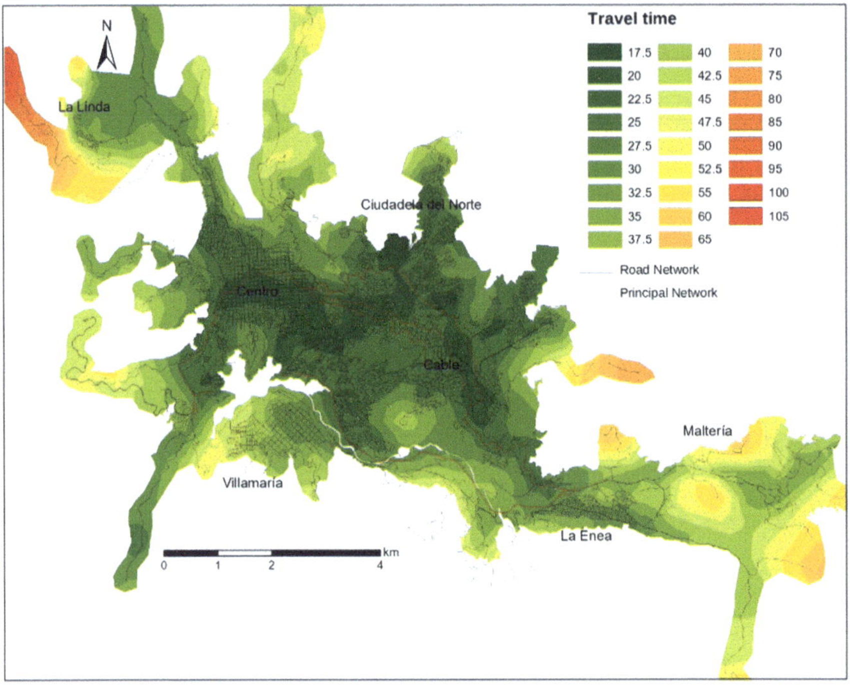

Fig. 2 Overall average accessibility in public mode

3.2 Integral Accessibility to Universities

The average accessibility conditions to the universities by public transport are shown in Fig. 4. It can be seen that the area that crosses Avenida Santander has the best access conditions, with average journey times of up to 8 min, a larger time when compared to private mode. Likewise, times are increased throughout the map, reaching extremes of 96 min.

In Villamaría, the times are high, up to 40 min, despite the existence of the aerial cable that reaches the municipality's square, which impacts the times. In this case, this mode of transport does not fulfill this task at a very high rate due to the location of the universities, far from the area of influence of the current aerial cable stations. Figure 5 shows that the best access conditions, in this mode of transport, are found in the higher strata and private mode. The lower strata have the lowest percentage coverage, barely reaching 20% coverage for travel times of up to 10 min, by which time the other strata already have coverage of over 50%, even reaching 89% for stratum five. Stratum 3 is in an intermediate range of coverage, being surpassed by strata 4, 5, and 6 in the vast majority of the analysis.

Figure 6 shows the accessibility results representing average travel times to the nearest university by private transport. The best accessibility conditions are found in the cable sector, and in general along Carrera 23 or Avenida Santander, with

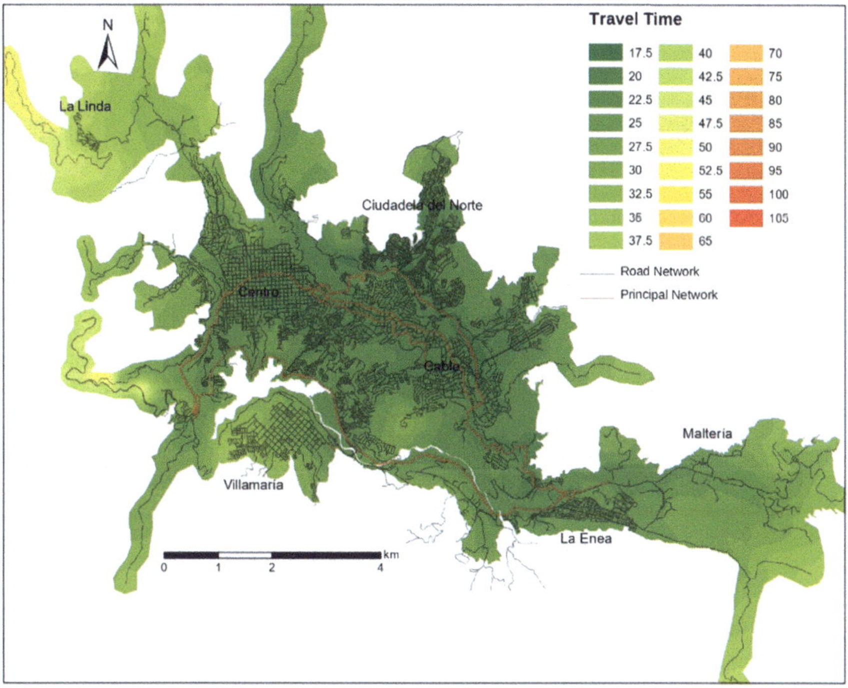

Fig. 3 Overall average accessibility in private mode

average travel times of less than 5 min. Other areas report average travel times of less than 5 min, such as a small sector of La Enea to the southeast and the exit to the neighboring municipality of Chinchiná to the west due to the presence of Sena and La Nubia Campus in La Enea sector and the Antonio Nariño University to the west.

On the other hand, the longest times are found in the northwest of the area studied, beyond La Linda neighborhood and in the innermost part of the urban area of the municipality of Villamaría, which has restricted access, so that their times have as a minimum basis the coverage in the connection sector. The maximum time to access higher education by private transport is more than half an hour, and Fig. 7 shows the percentages of accumulated percentage coverage differentiated by socioeconomic stratum. There is a clear difference in coverage between the low and high strata, with stratum five being almost fully covered in times of less than 8 min, while for the same time range strata one and two still do not exceed 20% coverage. Strata four and six are slightly below stratum five and obtain values close to 100% coverage in just over 12 min, while stratum three covers almost half of its population in less than 8 min. Its coverage is reduced and contemplates times of up to 28 min to cover 100%, being surpassed even by stratum one, whose population enjoys total coverage from 23 min, despite not exceeding 40% in curves of less than 10 min. Stratum two is the stratum with the most deficient coverage.

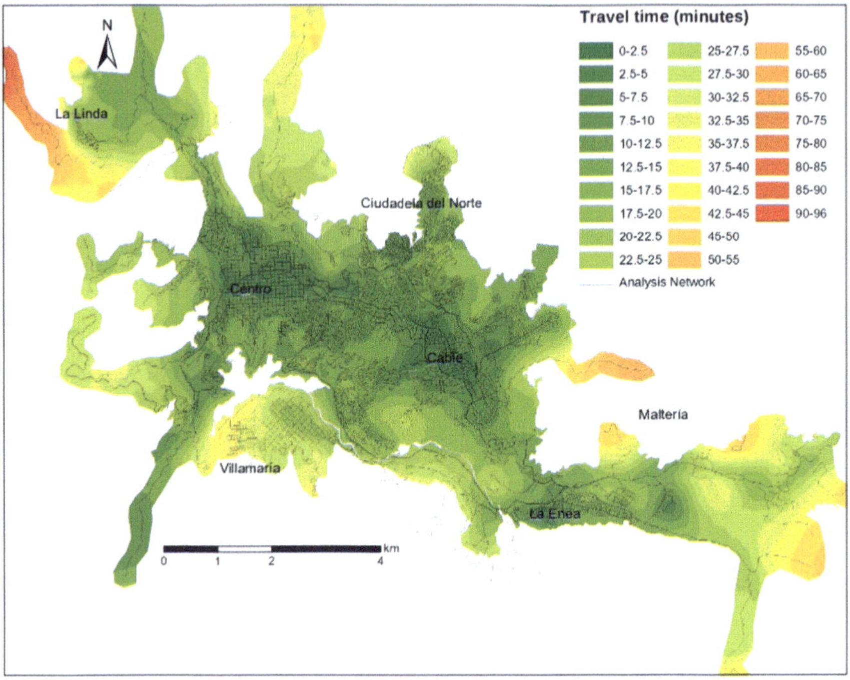

Fig. 4 Average accessibility to universities by public transport

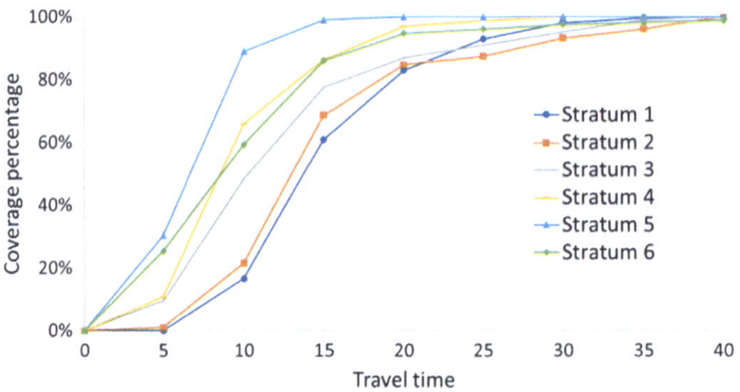

Fig. 5 Percentage of population coverage by public transport mode, by socioeconomic stratum

Figure 8 shows the result of comparing the average travel times of the average comprehensive accessibility conditions to the universities in public transport and private transport modes. The sectors in green are those in which it is more efficient to use public transport to reach a university institution, while the sectors in gray are those in which it is more efficient to reach a university institution by private transport. It can be seen that the sectors in green are disjointed, given the operational characteristics

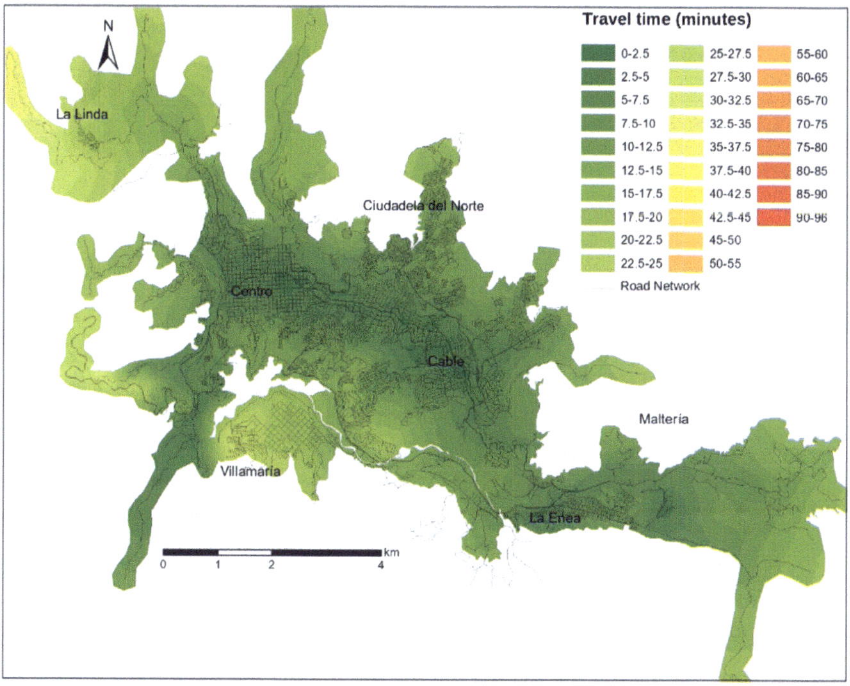

Fig. 6 Average accessibility to universities by private transport

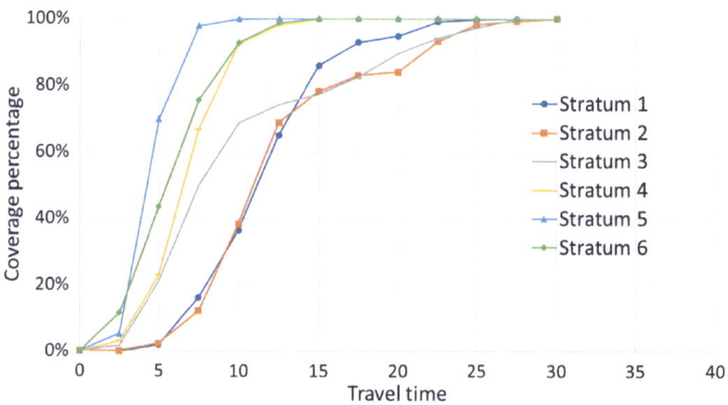

Fig. 7 Percentage of population coverage by private transport mode, by socioeconomic stratum

of both modes of transport and given the conditions of the infrastructure network, as well as the location of the 24 universities throughout the metropolitan area. In the east of the city, despite the existence of two of the largest universities (Universidad Nacional and SENA), their accessibility conditions are better in private mode, even though they are public universities. On the other hand, the sector of the University

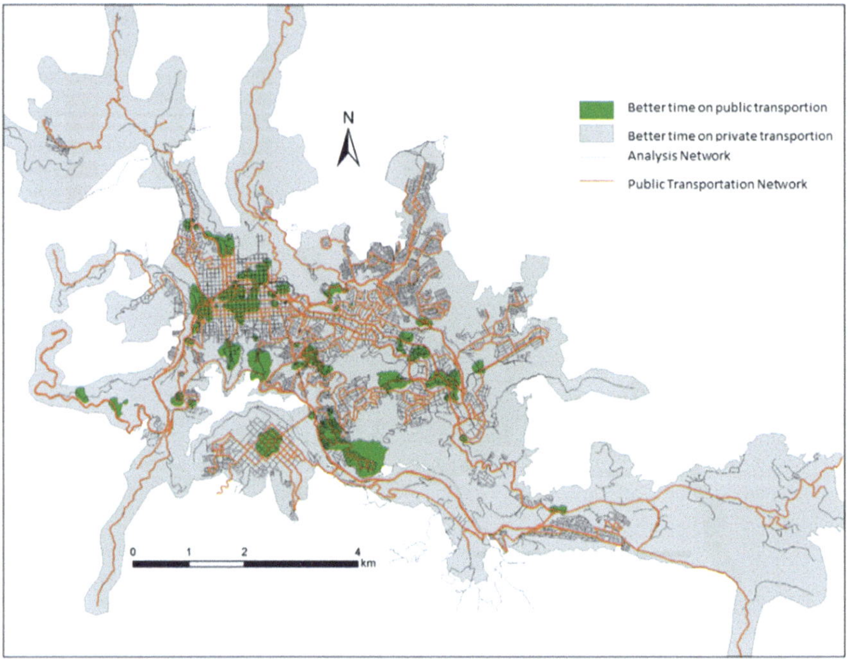

Fig. 8 Ratio between average travel times in public and private mode

of Manizales, which is private, has better conditions of access by public transport, which can be positive in the sense of offering alternative and sustainable modes to its students and professors, most of whom travel by private vehicle.

However, on analyzing the sectors where access to public transport would be best, it was found that only 11.4% of the population is in these sectors, which means that 89% of the population have private transport as the most efficient means of transport to reach the universities. This situation goes totally against the guidelines and directives defined in terms of sustainable mobility and equity, as the vast majority of university students do not have the economic capacity to acquire a vehicle, and the city should promote the use of public and sustainable modes of transport to those in the process of training and education.

4 Conclusions

The most unfavorable accessibility conditions correspond to higher education, due to its reduced number of establishments compared to the rest of the educational levels. Stratum 5 has the best percentage coverage when it comes to accessing university institutions due to its concentrated location in the central area of the city, bordering the main roads that provide quick access to and from the university, whether by public or private transport. The differences in accessibility when traveling by public and private transport are very noticeable in the outlying areas of the metropolitan area, but in the urban CBD, access times to the universities are very similar. Villamaría has strong connectivity drawbacks by having a single point of entry and exit of motor vehicles in connection with the capital of the department; these conditions its travel times and prevents it from quickly accessing geographically close equipment, such as the Antonio Nariño University, by forcing vehicles to cross the entire urban area to the main road connection. Such methodological instruments allow municipal administrations to establish clear routes of action to adjust public transport systems and to achieve, in some reasonable period, that the number of people enjoying better access conditions using public transport mode increases. Until then, people and, in this case, university students will be disadvantaged concerning the conditions of access to their training sites.

Acknowledgements This research is part of the call for the support of the Faculty of Engineering and Architecture research projects, Universidad Nacional de Colombia—Sede Manizales, 2019, through the project entitled: Methodological proposal for the calculation of an Index of Educational Equity at the level of basic, secondary, and higher education in the urban area of Manizales and Villamaría municipalities—Caldas. Hermes code: 45892. Thanks are due to the members of the Sustainable Mobility Research Group (GIMS) COL0011099201909021520.

References

1. Consejo Privado de Competitividad & Universidad del Rosario (2020) Índice De Competitividad De Ciudades 2020 [Cities competitiveness index 2020]
2. de Manizales A (2020) General information. Available: https://manizales.gov.co/informacion-general/
3. Hansen WG (1959) How accessibility shapes land use. J Am Inst Plan 25(2):73–76. https://doi.org/10.1080/01944365908978307
4. Morris JM, Dumble PL, Wigan MR (1979) Accessibility indicators for transport planning. Transp Res Part A Gen 13(2):91–109. https://doi.org/10.1016/0191-2607(79)90012-8
5. Dalvi MQ, Martin KM (1976) The measurement of accessibility: some preliminary results. Transportation 5(1):17–42. https://doi.org/10.1007/BF00165245
6. Geurs KT, van Wee B (2004) Accessibility evaluation of land-use and transport strategies: review and research directions. J Transp Geogr 12(2):127–140. https://doi.org/10.1016/j.jtrangeo.2003.10.005
7. Khan AA (1992) An integrated approach to measuring potential spatial access to health care services. Socioecon Plann Sci 26(4):275–287. https://doi.org/10.1016/0038-0121(92)90004-O
8. Joseph AE, Bantock PR (1982) Measuring potential physical accessibility to general practitioners in rural areas: a method and case study. Soc Sci Med 16(1):85–90. https://doi.org/10.1016/0277-9536(82)90428-2
9. Luo W, Wang F (2003) Measures of spatial accessibility to health care in a GIS environment: synthesis and a case study in the Chicago region. Environ Plann B Plann Des 30(6):865–884. https://doi.org/10.1068/b29120
10. Organización de las Naciones Unidas—ONU (2018) Objetivos de Desarrollo Sostenible 2018. Available at: http://www.undp.org/content/undp/es/home/sustainable-development-goals.html
11. Villanueba A (2010) Accesibilidad geográfica a los sistemas de salud y educación. Análisis espacial de las localidades de Necochea y Quequén [Geographic accessibility to health and

education systems. Spatial analysis of the localities of Necochea and Quequén]. Rev transp territ 2:136–157
12. Walsh S, Flannery D, Cullinan J (2015) Geographic accessibility to higher education on the island of Ireland. Irish Educ Stud 34(1):5–23. https://doi.org/10.1080/03323315.2015.1010302
13. Frenette M (2006) Too far to go on? Distance to school and university participation. Educ Econ 14(1):31–58. https://doi.org/10.1080/09645290500481865
14. Blanden J, Machin S (2004) Educational inequality and the expansion of UK higher education. Scott J Polit Econ 51(2):230–249. https://doi.org/10.1111/j.0036-9292.2004.00304.x
15. Gibbons S, Vignoles A (2012) Geography choice and participation in higher education in England. Reg Sci Urban Econ 42(1–2):98–113. https://doi.org/10.1016/j.regsciurbeco.2011.07.004
16. Sá C, Amado D, Justino E, Amaral A (2011) Higher education (related) choices in Portugal: joint decisions on institution type and leaving home. Stud High Educ 36(6):689–703. https://doi.org/10.1080/03075071003725343
17. Sá C, Florax RJ, Rietveld P (2006) Does accessibility to higher education matter? Choice behaviour of high school graduates in the Netherlands. Spat Econ Anal 1(2):155–174. https://doi.org/10.1080/17421770601009791
18. Spiess CK, Wrohlich K (2010) Does distance determine who attends a university in Germany? Econ Educ Rev 29(3):470–479. https://doi.org/10.1016/j.econedurev.2009.10.009
19. Montoya JA, Escobar DA, Younes C (2018) Coverage of educational centers from territorial accessibility in the municipality of Pitalito—Huila, Colombia. Contemp Eng Sci 11(61):3021–3033. https://doi.org/10.12988/ces.2018.87310
20. Avendaño NR (2012) La Equidad Espacial En La Distribución Del Servicio Educativo (1999–2009) [Spatial equity in the distribution of educational service (1999–2009)]. Universidad Nacional de Colombia, Bogotá
21. Younes C, Escobar DA, Holguín JM (2016) Equidad, accesibilidad y transporte. Aplicación explicativa mediante un análisis de accesibilidad al sector universitario de Manizales (Colombia) [Equity, accessibility and transportation. Explanatory application through an accessibility analysis to the university sector in Manizales (Colombia)]. Inf tecnol 27(3):107–118. https://doi.org/10.4067/S0718-07642016000300010
22. Escobar DA, Garcia FJ (2012) Territorial accessibility analysis as a key variable for diagnosis of urban mobility: a case study Manizales (Colombia). Procedia Soc Behav Sci 48:1385–1394. https://doi.org/10.1016/j.sbspro.2012.06.1114
23. Montoya JA (2019) Análisis de cobertura, mediante accesibilidad geográfica, para sistemas de paraderos de transporte público colectivo urbano [Coverage analysis, by means of geographic accessibility, for urban public transportation bus stop systems]. M.Sc. thesis, Universidad Nacional De Colombia Sede Manizales, Manizales
24. Giraldo R (2002) Introducción a la Geoestadística: Teoría y Aplicación [Introduction to geostatistics: theory and application]. Universidad Nacional de Colombia, Bogotá

Functioning Paid Parking Zone Before and During COVID-19 Pandemic—Case Study on the Example of a City in Poland

Agata Kurek and Elżbieta Macioszek

Abstract The constantly growing number of vehicles moving on the city's transport network causes traffic disruptions, lowering the level of road safety, as well as the problem of finding a vacant place to park the vehicle. The smallest number of parking spaces is in the city center area with many travel destinations. Therefore, more and more cities in Poland are introducing the paid parking zone (PPZ), which increases the availability of parking spaces for more drivers. The article aims to analyze the functioning of PPZ in Chorzów (Poland) before and during the COVID-19 pandemic. The analysis was carried out based on the number of purchased parking tickets in particular months, days of the week, and periods of the day in 2019 and 2020. It can be concluded that the COVID-19 pandemic changed the communication behavior in terms of PPZ use, based on the analyses. The short period for which parking tickets were purchased in 2019 and 2020 indicates that charging fees for parking at PPZ in Chorzów contribute to the choosing of a short parking time and thus increase the availability of parking spaces for more drivers.

Keywords City parking · Paid parking zone · COVID-19 pandemic

1 Introduction

Currently, transport network in cities is more and more cars. The increase in traffic volume causes not only traffic disruptions but also a reduction in the level of road safety and the problem of finding a vacant space to park the vehicle. The city center is usually the location of the largest number of daily travel destinations for residents. Therefore, these areas are often characterized by the smallest number of parking spaces. The search for vacant parking spaces leads to search traffic, which also

---

A. Kurek (✉) · E. Macioszek
Department of Transport Systems, Traffic Engineering and Logistics, Faculty of Transport and Aviation Engineering, Silesian University of Technology, Krasińskiego 8 Street, 40-019 Katowice, Poland
e-mail: agata.kurek@polsl.pl

E. Macioszek
e-mail: elzbieta.macioszek@polsl.pl

M. Petrović et al. (eds.), *The Science and Development of Transport—ZIRP 2021*,
https://doi.org/10.1007/978-3-030-97528-9_4

negatively affects the natural environment and causes frustration and irritation of drivers [1–3]. Therefore, more and more cities in the world, including Poland, are introducing the paid parking zone (PPZ). PPZ increases the availability of parking spaces for more drivers and can be part of the area traffic management system under ITS [1, 4].

The functioning of cities, the mobility of inhabitants, and the distribution of traffic on the transport network in the world changed with the spread of acute infectious disease of the respiratory system caused by the SARS-CoV-2—COVID-19 virus. The state of the epidemic was introduced on March 20, 2020, in Poland. The introduction of the epidemic was associated with the closure of services such as: restaurants, hotels, theaters, cinemas, museums, gyms, and the beauty industry. Teaching in schools and universities was carried out remotely. In many workplaces was recommended remote work, where it was possible. Some of the services were opened in the following weeks of the year, in compliance with the required sanitary regime, when the number of infected people began to decrease (in May 2020). In September 2020, most schools returned to teaching classes at school. From October 2020, laboratory, project, and exercises were conducted at universities, and lectures were conducted remotely. In October and November 2020, classes in schools and universities were again conducted remotely due to the constantly growing number of infected people. At the same time, most services were also closed, i.e., cinemas, theaters, museums, hotels, etc. [5].

The article aims to analyze the functioning of PPZ in Chorzów (Poland) before and during the COVID-19 pandemic. The number of purchased parking tickets in particular months, days of the week, and periods of the day in 2019 and 2020 was analyzed. This analysis allowed determining the differences in the use of PPZ in Chorzów in 2019 and 2020. The article consists of four parts. After the introduction and literature review on the functioning of parking in the city, characteristic of the research area and conducted research were presented. This section covers the characteristics of the research area, collected data, and analyses presented in the article. An analysis of the functioning of the PPZ in Chorzów (Poland) in 2019 and 2020 is presented in the next, fourth part. The article ends with conclusions resulting from the conducted analysis.

2 Literature Review

The studies presented in the literature include analyses of the use of parking spaces. They apply to different parking types, e.g., P&R parking [6, 7], parking at shopping centers, parking at workplaces, parking at universities [8], and parking covered by PPZ [9, 10].

A paper covering the subject of paid parking is, e.g., [11], which presents the use of parking space at the shopping center before and after the introduction of parking fees. The use of the car when traveling to the shopping center decreased, as well as the share of stopping vehicles in prohibited places after the introduction of obligatory fees for using the parking. In turn, in [12], the impact of introducing parking charges on the time needed to find a parking space was compared and hence the impact on traffic caused by looking for a parking space. The research results show that the introduction of a parking fee resulted in a reduction of the parking time and a reduction in the use of parking space. Thanks to this, the chance of finding a vacant parking space increased. Additionally, the traffic generated by searching for a parking space decreased. Whereas in the work [13], a model was proposed for predicting the impact of introducing a parking fee or changing the amount of the fee and introducing a parking time limitation on the choice of travel mode. The results show that the pricing of parking the most affects commuters and car use. In turn, limiting the parking time affects the choice of parking. In the next work, Nourinejad and Roorda [14] presented the results of research on the impact of the hourly fee on the demand for parking spaces. The results indicate that the amount of the parking charge may decrease or increase the demand for parking spaces, which depends on the flexibility of parking time to the price per hour of parking. The studies related to the determination of the appropriate parking rate were carried out by the authors of Simićević et al. [15]. The most important factor in parking selection is the shortening of the duration of searching for a parking space based on the results of the polynomial logit model presented in the paper. In turn, Zakharov et al. [16] presented studies on the impact of PPZ on the use of means of transport and the effectiveness of traffic management and the natural environment. They proposed a factor for the environmental sustainability of the transport system. All of the above-mentioned works indicate that introducing paid parking causes changes in communication behavior in a given area. Moreover, Kozłowski et al. [17] conducted tests to verify the correct operation of the control system for parking fees paid at PPZ in Warsaw. The authors conclude from the research that the functionality of the control system is satisfactory.

The process of collecting data in the field for analyses in the area of traffic engineering, road transport (including analyses of the operation of parking) is a time-consuming task. Therefore, in practice, many different techniques are supporting this process. The most convenient situation is having data from automatic measurements. Amato et al. [18] present the use of a visual vehicle presence detection solution based on convolutional neural networks (CNN). In turn, the paper [19] presents the use of wireless sensor networks (WSN) to detect vehicles at parking spaces. WSN consists of wireless sensors located on parking spaces, sending information about the occupancy of parking spaces to the embedded web server. In turn, Dalarmelina et al. [20] present the possibility of collecting data from parking based on optical character recognition (OCR). OCR enables automatic recognition of license plates from the video image from cameras installed in the city, which are connected to the Internet.

The spread of the SARS-CoV-2 virus has changed many areas of people's lives. To minimize the probability of becoming ill, inhabitants limited their travels to those necessary for everyday matters. This contributed to changes in the distribution of traffic on the city's transport network. The articles in world literature investigate the impact of a pandemic on changes in the transport system of cities. For example, research conducted by Aloi et al. [21] indicates that in March 2020, there was a decrease in car traffic by 76% compared to the period before the pandemic and a decrease in the number of public transport users by as much as 93%. The impact of the SARS-CoV-2 virus on the use of public transport is also presented in the paper Sahraei et al. [22]. In March 2020, the use of public transport in travel in Ankara and Istanbul decreases by 80%. Sahraei et al. also analyzed air quality, which in the city of Istanbul improved by 47%, while in Ankara by 9%. In turn, the impact of the pandemic on the city's transport system was studied by Gao et al. [23]. These studies show that the number of parked vehicles in the city decreased by about 70% in March 2020 compared to the corresponding period in 2019. Delot and Ilarri [24] proposed an innovative approach allowing for the identification of such parking places, which—by maintaining an appropriate social distance—can contribute to the prevention of the spread of the SARS-CoV-2 virus.

3 Characteristic of the Research Area and Conducted Research

The paid parking zone in Chorzów (Poland) is divided into two subzones: A and B. Subzone A includes 23 streets (about 1000 parking spaces), while subzone B—18 streets (about 300 parking spaces). Figure 1 shows the area covered by the PPZ in Chorzów with the division into subzones.

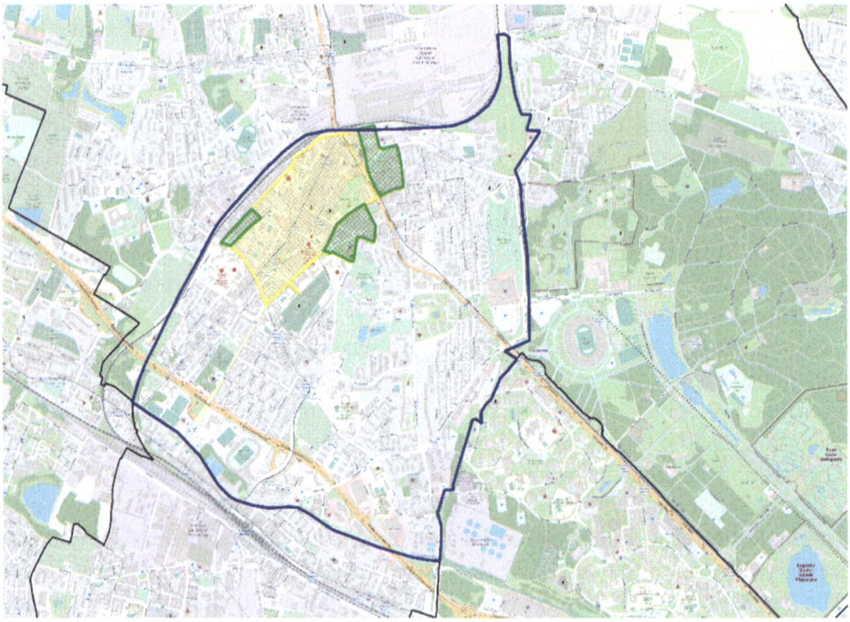

Fig. 1 Area of PPZ in Chorzów divided into subzone A (yellow), subzone B (green), and city center location (blue). *Source* Own based on [25]

Table 1 Parking fee in the subzones of PPZ in Chorzów

Period	Subzone	
	A	B
The minimum parking time—15 min	1.00 PLN (~0.22 EUR)	0.50 PLN (~0.11 EUR)
The first, fourth, and fifth hour of parking	3.00 PLN (~0.66 EUR)	2.00 PLN (~0.44 EUR)
The second hour of parking	3.40 PLN (~0.75 EUR)	2.20 PLN (~0.48 EUR)
The third hour of parking	3.90 PLN (~0.86 EUR)	2.40 PLN (~0.53 EUR)
The subscription fee for five hours or more on a given day	16.30 PLN (~3.59 EUR)	10.60 PLN (~2.33 EUR)

Source [27]

According to [26], parking fees at PPZ in Poland cannot be charged on weekend days and holidays. In Chorzów, the parking fee is charged on working days between 10.00 and 18.00 (based on the Resolution of the Chorzów City Council: [27]). The fees are presented in Table 1. The fee does not include:

- Disabled people and people transporting disabled people with a parking card;
- Users of two-wheeled vehicles;
- Users of permanently marked vehicles of the Municipal Police in Chorzów.

Inhabitants living in the streets covered by the PPZ may pay a flat fee for parking only one car. The fee is PLN 200 per year [27].

The user of PPZ in Chorzów, who does not pay the parking fee, must pay an additional fee of 50 PLN (~11 EUR).

There are several ways to pay the parking fee in PPZ. The fee can be paid in the machines: MBS and Silesian Public Service Card (SKUP) or using the mobile applications: Pango and mobiParking. In the case of the MBS machine, the payment can be made in cash or with a city card and in the SKUP machine—in cash or with a SKUP card. In the mobile application, the payment can be made using a mobile card or transfer.

The PPZ in Chorzów is not equipped with parking space monitoring devices included in the Intelligent Transport System. Therefore, the analysis of the functioning of the PPZ zone in 2019 and 2020 was made based on the number of purchased parking tickets authorizing to park in the PPZ. The data was obtained from the Municipal Management of Streets and Bridges in Chorzów and included information on purchased tickets using four tools (MBS ticket machine, SKUP ticket machine, Pango mobile application, mobiParking mobile application) on each day in 2019 and 2020. These data allowed for the following analyses for 2019 and 2020:

- The share of purchased parking tickets with the use of particular tools;
- The share of purchased parking tickets with the use of particular payment methods;
- The share of the number of purchased parking tickets for one parking space in subzones A and B;

- The number of purchased parking tickets on working days between 10:00 and 18:00 in particular months of the year;
- The average number of purchased parking tickets on working days between 10:00 and 18:00 in particular months;
- The share of parking tickets purchased outside the period of validity of the PPZ in Chorzów in particular months;
- The average number of parking tickets purchased between 10:00 and 18:00 on particular working days and the average number of parking tickets purchased on Saturdays and Sundays;
- The average number of purchased parking tickets in particular hours on working days;
- Distribution of estimated parking time at PPZ in Chorzów on an average working day.

4 Analysis of the Functioning of Paid Parking Zone in Chorzów Before and During COVID-19 Pandemic

Figure 2 presents the share of purchased parking tickets with the use of particular tools in 2019 and 2020. In 2019, most tickets were purchased from the MBS ticket machine—almost 53%, then from SKUP ticket machine—over 43%. More than 4% of parking tickets were purchased using mobile applications (mobiParking—3.50%, Pango—0.54%). The use of the mobile application in paying parking fees in PPZ in Chorzów increased in 2020 compared to 2019—in total over 6%. In addition, more parking tickets in 2020 were purchased with the SKUP ticket machine (an increase by over 13% compared to 2019) than for the MBS ticket machine (a decrease by over 15% compared to 2019).

Figure 3 shows the share of purchased parking tickets using particular payment methods in 2019 and 2020. In 2019, most parking tickets were paid in cash—over 94%. Parking tickets paid with mobile payments were 4.19%, SKUP card—1.15%, and city card—0.15%. The increase in the use of mobile applications in paying

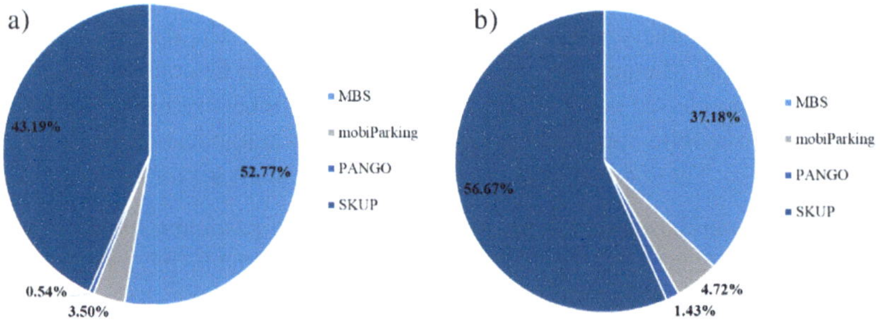

Fig. 2 Share of purchased parking tickets with the use of particular tools in **a** 2019; **b** 2020

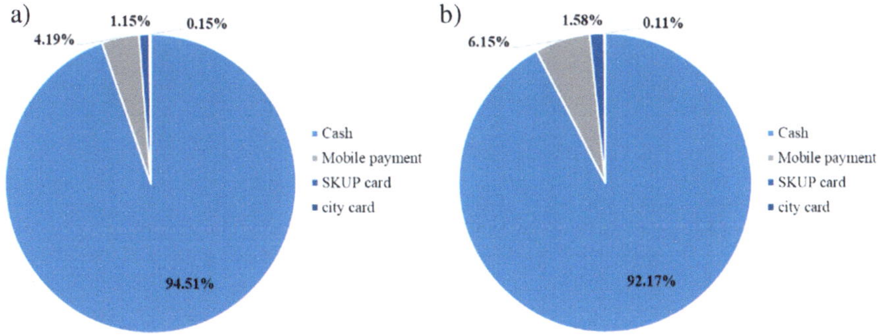

Fig. 3 Share of purchased parking tickets with the use of particular payment methods in **a** 2019; **b** 2020

parking fees in PPZ in Chorzów in 2020 compared to 2019 is reflected in the increase in the use of mobile payments (an increase by almost 2%). Cash payments decreased (less than 2%) for parking at PPZ in Chorzów in 2020 compared to 2019. The form of payment is related to the payment tool used. Almost 96% in 2019 and almost 94% in 2020 parking fees were paid in ticket machines, where it is possible to pay in cash, with the resident's card, and with the SKUP card.

Figure 4 presents the share of the number of purchased parking tickets per parking space in a particular subzone in 2019 and 2020. In subzone B, more parking tickets were purchased for one parking space (58.14%) than in subzone A in both 2019 and 2020, although there are more parking spaces in subzone A. In 2020, this share decreased by over 5% compared to 2019. The increase in the number of parked vehicles in subzone A maybe because this subzone is located closer to the city center, where there are more destinations (Fig. 1). It can be concluded that users of PPZ in Chorzów preferred to park closer to the destination, despite the higher parking fee, to minimize contact with other people, due to the COVID-19 pandemic.

Figure 5 shows the average number of purchased parking tickets on working days

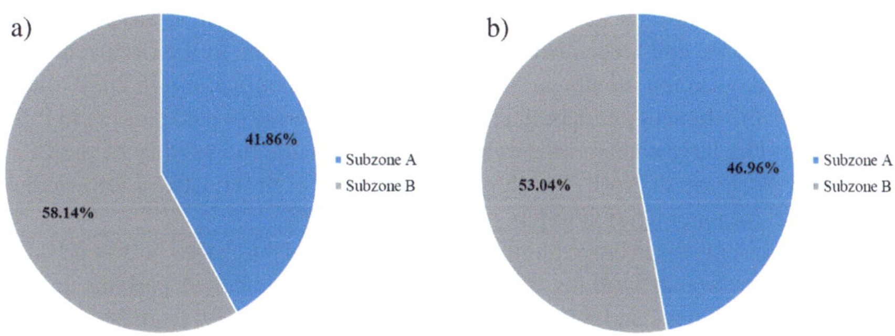

Fig. 4 Share of the number of purchased parking tickets per parking space in subzones A and B in **a** 2019; **b** 2020

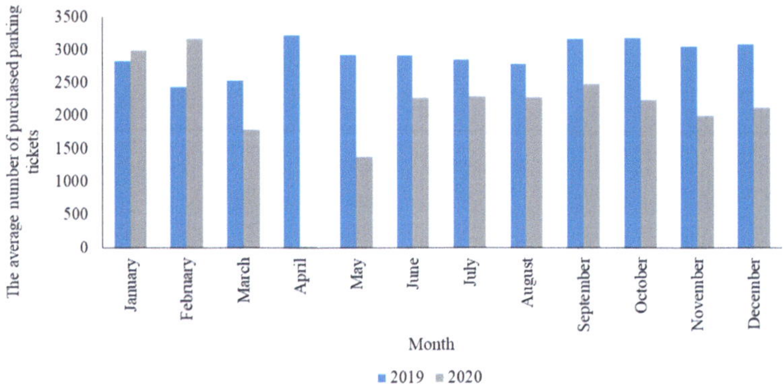

Fig. 5 Average number of purchased parking tickets on working days between 10:00 and 18:00 in particular months in 2019 and 2020

from 10:00 to 18:00 in particular months of 2019 and 2020. In 2019, the largest number of parking tickets was purchased in April, September, October, November, and December, however the least in February and March. At the beginning of 2020 (January and February), there were more parking tickets purchased compared to the same period in 2019. From March 2020, there was a decrease in the number of purchased parking tickets compared to 2019. In April, fees for parking at PPZ in Chorzów were not collected to try to limit the spread of the SARS-CoV-2 virus. Users who paid in April 2020 for parking at PPZ in Chorzów probably did not have information about the lack of need to pay parking fees. In September 2020, a slight increase in the number of purchased parking tickets can be observed. This could be related to the end of the holiday period and the lifting of restrictions related to COVID-19. From October 2020, another decrease in purchased parking tickets has been observed, which may be caused by the reintroduction of restrictions in Poland due to the increase in the number of people infected with the SARS-CoV-2 virus.

Table 2 presents the number and share of purchased parking tickets outside the period of validity of the PPZ in Chorzów in particular months of 2019 and 2020. The largest number of parking tickets outside the validity period of the PPZ was purchased in June, November, and December in 2019. This may be due to the occurrence of non-working days in the months in which payment is not obligatory. In 2020, there an increase in the share of purchased tickets beyond the period of paid parking in PPZ in Chorzów. The largest share of purchased tickets outside the validity of the PPZ in Chorzów occurred in April, which probably results from the lack of knowledge about the lack of an obligation to pay parking fees at PPZ.

Figure 6 presents the average number of parking tickets purchased between 10:00 and 18:00 on particular working days and the average number of parking tickets purchased on Saturdays and Sundays in 2019 and 2020. In 2019, from Monday to Friday, the number of purchased parking tickets is at the level of 3000. Slightly more purchased parking tickets can be observed on Mondays and Thursdays (respectively, 3012 and 3030 parking tickets purchased). In 2020, on working days, the number of

Table 2 Share of parking tickets purchased outside the period of validity of the PPZ in Chorzów in particular months in 2019 and 2020

Month	The share of purchased tickets outside the period of the PPZ in Chorzów [tickets]	
	2019	2020
January	8475 (13.61%)	9363 (14.89%)
February	6571 (13.46%)	9923 (15.66%)
March	7880 (14.76%)	6438 (16.40%)
April	9740 (14.37%)	110 (22.27%)
May	9072 (14.73%)	4062 (14.77%)
June	8352 (15.05%)	7581 (15.89%)
July	9344 (14.21%)	8431 (15.94%)
August	8337 (14.24%)	7726 (16.08%)
September	9737 (14.59%)	8649 (15.81%)
October	10,522 (14.39%)	8107 (16.43%)
November	9016 (15.55%)	6536 (16.36%)
December	9303 (15.04%)	7320 (15.61%)

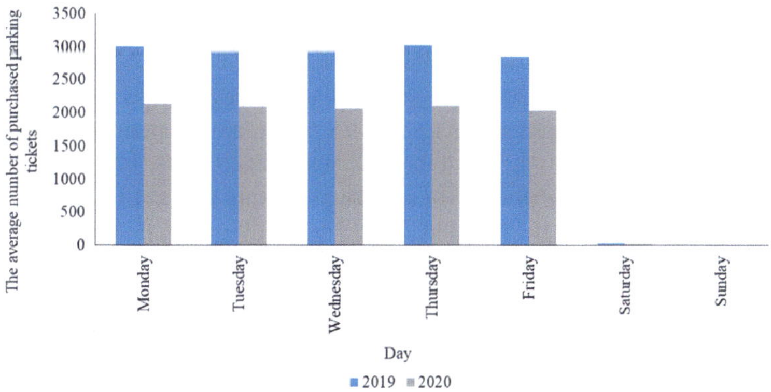

Fig. 6 Average number of parking tickets purchased between 10:00 and 18:00 on particular working days and the average number of parking tickets purchased on Saturdays and Sundays in 2019 and 2020

purchased parking tickets is also at a similar level and fluctuates around 2000. Also on Mondays and Thursdays, slightly more purchased parking tickets can be observed (respectively, on average 2138 and 2106 parking tickets purchased). On Saturdays and Sundays, the number of parking tickets purchased is small (respectively 33 and 10 in 2019 and 16 and 5 parking tickets in 2020). This is due to the lack of an obligation to pay parking fees at PPZ in Chorzów on weekend days.

Figure 7 presents the average number of purchased parking tickets in particular hours on working days in 2019 and 2020. The most purchased tickets during the validity period of the PPZ in Chorzów were purchased between 10:00 and 13:00,

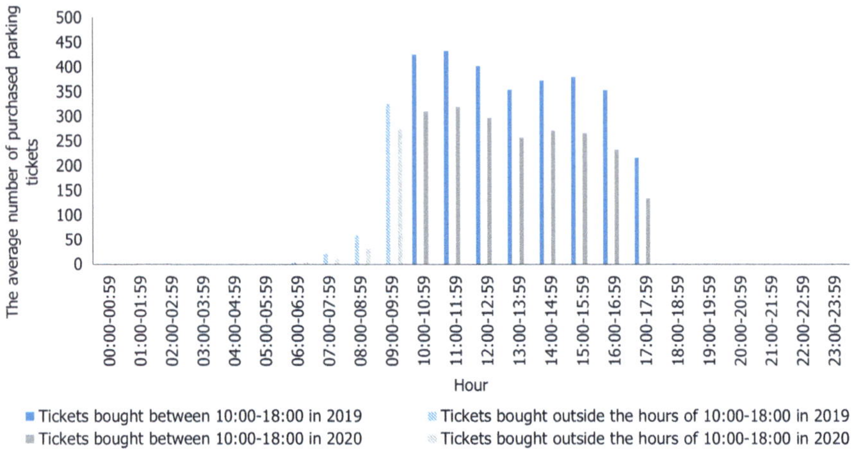

Fig. 7 Average number of purchased parking tickets in particular hours on working days in 2019 and 2020

while the least between 17:00 and 18:00 in 2019 and 2020. Outside the period of validity of PPZ in Chorzów, users also paid parking fees. The largest number of purchased parking tickets outside the period of validity of the PPZ in Chorzów was between 09:00 and 10:00 in 2019 and 2020. This may result from the fact that the driver, upon arrival at the parking space, pays a fee before the period of validity of the PPZ, which is charged only when the PPZ starts functioning (i.e., from 10:00).

Figure 8 presents the distribution of estimated parking times at PPZ in Chorzów on an average working day in 2019 and 2020. The user buying a parking ticket from the ticket machine must specify for how long they plan to park. In the case of the mobile application, the user can purchase a ticket for a predetermined time or turn on the parking time when parking starts and turn off the parking time when it ends.

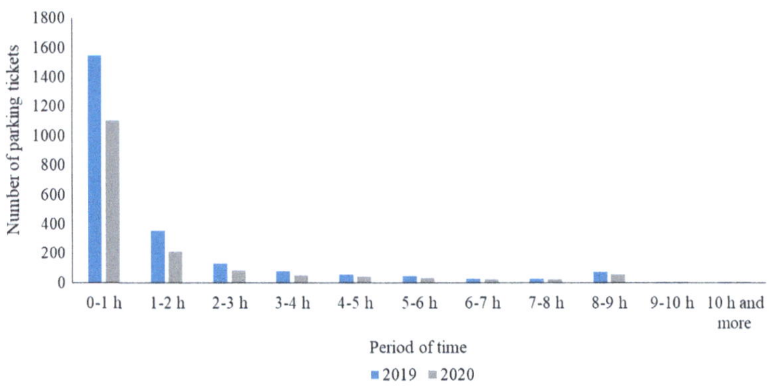

Fig. 8 Distribution of estimated parking time at PPZ in Chorzów on an average working day in 2019 and 2020

Most people left their vehicles for up to one hour in 2019 and 2020 and the least for a period of 2–8 h. It can therefore be concluded that charging fees for parking at PPZ in Chorzów contributes to the reduction of the parking time and thus increases the availability of parking spaces for a larger number of drivers.

5 Discussion and Conclusions

The article aimed to analyze the functioning of PPZ in Chorzów (Poland) before and during the COVID-19 pandemic. The number of purchased parking tickets in particular months, days of the week, and periods of the day in 2019 and 2020 was analyzed based on the obtained data. The comparisons made in the article allowed for the formulation of the following conclusions:

- Most tickets were purchased from ticket machines (95.96% in 2019 and 93.85% in 2020). In the case of mobile applications, it was a total of 4.04% in 2019 and 6.15% of parking tickets purchased in 2020;
- Most parking tickets were paid in cash (over 94% in 2019 and 92.17% in 2020), than using mobile payments (4.19% in 2019 and 6.15% in 2020) and SKUP card (1.15% in 2019 and 1.58%in 2020) and city card (0.15% in 2019 and 0.11% in 2020);
- The increase in the share of mobile payments in 2020 compared to 2019 may be due to users' concerns about SARS-CoV-2 infection through the touch of ticket machines;
- In subzone B, more parking tickets per parking space were purchased (58.14% in 2019 and 53.04% in 2020) than in subzone A, although there are more parking spaces in subzone A;
- The largest number of parking tickets was purchased in April, September, October, November, and December, and the least in February and March in 2019. However, in the case of 2020, the largest number of parking tickets was purchased before the COVID-19 pandemic—in January and February. During the COVID-19 pandemic, in 2020, most parking tickets were purchased in September;
- The share of purchased parking tickets outside the validity period of the PPZ in Chorzów in 2019 ranges from 13.46 to 15.55% in 2019 and 14.77 to 22.27% in 2020;
- On working days, the number of purchased parking tickets is at the level of 3000 in 2019, while in 2020—about 2000. A slightly more purchased parking tickets can be observed on Mondays and Thursdays (respectively 3012 and 3030 in 2019; respectively 2106 and 2138 in 2020). On Saturdays and Sundays, the number of purchased parking tickets is small (respectively 33 and 10 in 2019; respectively 16 and 5 in 2020). This is due to the lack of an obligation to pay parking fees at PPZ in Chorzów on weekend days;

- The most purchased tickets during the period of validity of the PPZ in Chorzów were purchased between 10:00 and 13:00, while the least between 17:00 and 18:00 in 2019 and 2020;
- The most purchased parking tickets outside the period of validity of the PPZ in Chorzów was between 9:00 and 10:00 in 2019 and 2020;
- Although there is no obligation to pay fees on weekends and non-working days, some users have purchased parking tickets. The same is for working days outside the hours of 10.00–18.00. In some cases, this may be because if someone parks the car before the time when the PPZ in Chorzów applies, they make a payment so as not to return to the ticket machine (if they do not use the mobile application). In other cases, it may result from the lack of attention to information by users;
- Most people left their vehicle for a period of up to 1 h in 2019 and 2020, and the least for a period of 2–8 h. It can therefore be concluded that charging fees for parking at PPZ in Chorzów contributes to the choosing of a short stopping time, and thus increases the availability of parking spaces for a larger number of drivers;
- The results of the analyses in the article indicate differences in the use of PPZ in Chorzów in 2020 compared to 2019. What could be caused by the COVID-19 pandemic and the reduction in the number of city travel and behavior to limit contact with other people.

Further research is planned based on the obtained results. It is planned to conduct research on the use of PPZ in Chorzów in subsequent waves of the COVID-19 pandemic and after the stabilization of the pandemic situation. The COVID-19 pandemic has given many people the opportunity to work remotely, which has reduced the number of daily trips. Some companies still work remotely, despite the stabilization of the pandemic situation. Therefore, the results of future research will reveal changes in communication behavior.

Acknowledgements The present research has been financed from the "Excellence Initiative—Research University" program implemented at the Silesian University of Technology, 2021, grant number: 12/040/SDU/10-22-01.

References

1. Shoup D (2006) Cruising for parking. Transp Policy 13(6):479–486
2. Duda-Wiertel U, Szarata A (2019) The analysis of transport-related behaviours of drivers in highly occupied paid parking zones. Adv Transp Stud 47:35–48
3. Arnott R, Williams P (2017) Cruising for parking around a circle. Transp Res Part B Methodol 104:357–375
4. Litman T (2005) Parking management best practices. American Planning Association, New York
5. Service of the Republic of Poland. Available online: https://www.gov.pl/. Accessed 20 Apr 2021

6. Macioszek E, Kurek A (2020) The use of a park and ride system—a case study based on the City of Cracow (Poland). Energies 13(13):1–26
7. Macioszek E, Kurek A (2021) The analysis of the factors determining the choice of park and ride facility using a multinomial logit model. Energies 14(1):1–33
8. Jedynak Z (2015) Ocena dostępności systemu parkingowego Politechniki Rzeszowskiej. Mod Manag Rev 20(22):59–68
9. Duda-Wiertel U (2017) Pomiar parkowania z wykorzystaniem zapisu wideo. Transp Miejski Reg 3:5–10
10. Jamroz K, Oskarbski J, Gumińska L, Zarembski J (2015) Wyniki badań parkingowych w obszarze śródmieścia Gdyni. In: Zeszyty Naukowo-Techniczne Stowarzyszenia Inżynierów i Techników Komunikacji w Krakowie. Seria: Materiały Konferencyjne, vol 1(105), pp 27–37
11. Au G, Young W (2016) The impact of the introduction of paid parking at a local retail precinct: a case study of the Yarraville Village paid parking scheme. In: Proceedings of the 27th ARRB conference, Melbourne, Victoria, 16–18 Nov 2016, pp 1–17
12. Cats O, Zhang C, Nissan A (2015) Empirical evaluation of an on-street parking pricing scheme in the city center. In: 94th annual meeting of proceedings of the 94th transportation research board annual meeting, p 29
13. Simićević J, Vukanović S, Milosavljević N (2013) The effect of parking charges and time limit to car usage and parking behaviour. Transp Policy 30:125–131
14. Nourinejad M, Roorda MJ (2017) Impact of hourly parking pricing on travel demand. Transp Res Part A Policy Pract 98:28–45
15. Simićević J, Milosavljević N, Maletić G (2012) Influence of parking price on parking garage users' behaviour. PROMET-Traff Transp 24(5):413–423
16. Zakharov D, Fadyushin A, Chainikov D (2020) Changes in the environmental sustainability of the urban transport system when introducing paid parking for private vehicles. Resources 9(9):1–17
17. Kozłowski M, Czerepicki A, Jaskowski P, Aniszewski K (2021) Analysis of the system of controlling paid parking zones. Sustainability 13(8):1–12
18. Amato G, Carrara F, Falchi F, Gennaro C, Meghini C, Vairo C (2017) Deep learning for decentralized parking lot occupancy detection. Expert Syst Appl 72:327–334
19. Yang J, Portilla J, Riesgo T (2012) Smart parking service based on wireless sensor networks. In: Proceedings of the 38th annual conference on IEEE industrial electronics society, Montreal, QC, 25–28 Oct 2012, pp 6029–6034
20. Dalarmelina NDV, Teixeira MA, Meneguette RI (2020) A real-time automatic plate recognition system based on optical character recognition and wireless sensor networks for ITS. Sensors 20(1):1–13
21. Aloi A, Alonso B, Benavente J, Cordera R, Echániz E, González F, Ladisa C, Lezama-Romanelli R, López-Parra Á, Mazzei V, Perrucci L, Prieto-Quintana D, Rodríguez A, Sañudo R (2020) Effects of the COVID-19 lockdown on urban mobility: empirical evidence from the city of Santander (Spain). Sustainability 12(9):1–18
22. Sahraei MA, Kuşkapan E, Çodur MY (2021) Impact of Covid-19 on public transportation usage and ambient air quality in Turkey. PROMET-Traff Transp 33(2):179–191
23. Gao J, Bernardes SD, Bian Z, Ozbay K, Iyer S (2020) Initial impacts of COVID-19 on transportation systems: a case study of the US epicenter, the New York metropolitan area. Available online: https://arxiv.org/abs/2010.01168
24. Delot T, Ilarri S (2020) Let my car alone: parking strategies with social-distance preservation in the age of COVID-19. Procedia Comput Sci 177:143–150
25. Open street map. Available Online: https://www.openstreetmap.org/#map=7/51.992/19.512
26. Ministry of Infrastructure, Act of March 21, 1985 on public roads, with subsequent changes
27. Municipal Management of Streets and Bridges in Chorzów. Available online: https://www.mzuim.Chorzów.eu/

Speed Transition Matrix Feature Extraction for Traffic State Estimation Using Machine Learning Algorithms

Leo Tišljarić, Filip Ribić, Željko Majstorović, and Tonči Carić

Abstract Feature extraction is a crucial part of data preparation when using machine learning algorithms, especially for emerging datasets. The speed transition matrix (STM) emerged only recently as a traffic data modeling technique. In this paper, key features from STMs are extracted and proposed for the purpose of traffic state estimation. This step simplifies the learning process and the interpretability of the results obtained when estimating the traffic state using the STMs. Using the proposed features, traffic state is estimated for the most crucial road segments in the City of Zagreb, Croatia. The method is evaluated on some of the most used machine learning algorithms, with the highest accuracy value obtained with decision tree and random forest algorithms.

Keywords Speed transition matrix · Machine learning · Feature extraction · Traffic state estimation

1 Introduction

Traffic state estimation is a vital research topic for transport engineers, especially when dealing with congestion estimation or traffic incident detection. Traditionally, traffic state estimation studies use theoretical mathematical models to describe the observed traffic flows. The main criticism of these kinds of traffic modeling is received regarding the questionable ability for applications in real-world scenarios. On the other hand, data-driven models emerged due to the deployment of various traffic sensors [1] and increased development of data processing techniques that followed after an exponential increase of computing power in the recent decade [2, 16].

The speed transition matrix (STM) emerged only recently as a novel traffic data modeling technique. It finds its usages across many fields that are part of intelligent transport systems (ITS) [7], such as providing route or traffic information [13] or detection of road traffic anomalies [14, 15]. The main advantage of STM is that it

L. Tišljarić (✉) · F. Ribić · Ž. Majstorović · T. Carić
Faculty of Transport and Traffic Sciences, University of Zagreb, Vukelićeva 4, 10000 Zagreb, Croatia
e-mail: ltisljaric@fpz.unizg.hr

M. Petrović et al. (eds.), *The Science and Development of Transport—ZIRP 2021*,
https://doi.org/10.1007/978-3-030-97528-9_5

does not suffer from the large deviations in the dataset as data are not aggregated in narrow time intervals. Moreover, STM can be processed as a classical matrix or as a traffic image [6]. In both cases, it is suitable for some advanced algorithms like tensor decomposition or convolutional neural networks (CNNs).

There are two most dominant approaches for feature extraction from road traffic data in the literature: (i) utilization of information regarding dependencies in traffic flows provided by a researcher and (ii) automatic extraction of features using deep learning or machine learning approaches [10].

Authors in Ermagun and Levinson [3] introduced a network weight matrix that utilizes traffic network graph characteristics. They combined features related to graph characteristics like betweenness centrality and traffic characteristics by weighting the graph with traffic volume. In Nguyen et al. [9], the authors proposed extracting visual features from traffic images that represent speeds on the observed highways. The feature extraction was done using point-based methods to extract local features (corners, blobs, etc.) and area-based features for more high-level features (shapes or areas). Authors in Xu et al. [17] extracted spatio-temporal traffic-related features using locality sensitive discriminant analysis (LSDA) with the goal of dimensionality reduction. CNN is the most used deep learning method for automatic feature extraction suited for the analysis of datasets represented as images. Authors in Ma et al. [6] used CNNs for automatic feature extraction with the goal of short-term speed prediction on urban roads. In Yu et al. [18], authors proposed combining long short-term memory (LSTM) and CNN for automatic feature extraction from traffic images. The main goal was long-term traffic state prediction by considering spatial and temporal components of the traffic network.

In this paper, we propose a method for extracting some of the most notable features from STM to enable simple clustering machine learning algorithms. This approach provides a more straightforward learning process, more interpretable results regarding the usage of more interpretable machine learning algorithms, and better results when a smaller amount of data are used as those algorithms require less data to train when compared to deep learning algorithms. The main advantage of the proposed feature extraction method is combining visual- and traffic-related features.

Contributions of this paper are as follows: (i) method for feature extraction from the STM traffic data model, (ii) extracted set of features that can be used for describing the STM, (iii) evaluated results on most used machine learning algorithms using the large global navigation satellite system (GNSS) dataset. Used code for this research is publicly available on the GitHub repository [12].

The rest of the paper is organized as follows. In Sect. 2, the methodology overview is given. Section 3 represents data and the obtained results for the feature extraction method evaluated on different machine learning algorithms. Finally, Sect. 4 gives final remarks, conclusions, and future work recommendations.

2 Methodology

This paper presents a methodology for extracting the features from the STM by following steps represented in Fig. 1: (i) STM computation from raw GNSS data, (ii) extraction of visual- and traffic-related features, and (iii) evaluation using the most used machine learning algorithms.

2.1 Speed Transition Matrix

Concept of the STM origins from a transition matrix explained in the Markov chain theory, where every matrix cell represents the probability of transition between two states of the system. The STM can be defined as a transition matrix that shows the probability of vehicle speed transition between two consecutive links in the road network, observed in the time interval Δt, where speed is computed as a harmonic speed.

Figure 2 represents the example of the STM construction and visualization of two examples, normal traffic pattern and congested traffic pattern. Example in Fig. 2a illustrates two transitions labeled with red and blue colors, where c and b links represent the origin and f and g links represent the destination links of the observed transitions. Then, the STM is constructed as follows: (i) for each vehicle that was traveling through observed transition, harmonic speed on the origin and the destination link is computed, (ii) all speeds are represented by the relative speed (0–100%), relative to the speed limit on the observed link, (iii) all speed transitions between origin and destination links are counted within specified time interval Δt and placed into the matrix, and (iv) counts are transformed into the probability by dividing each cell with the sum of all counts.

The STM can be defined as a matrix X as follows:

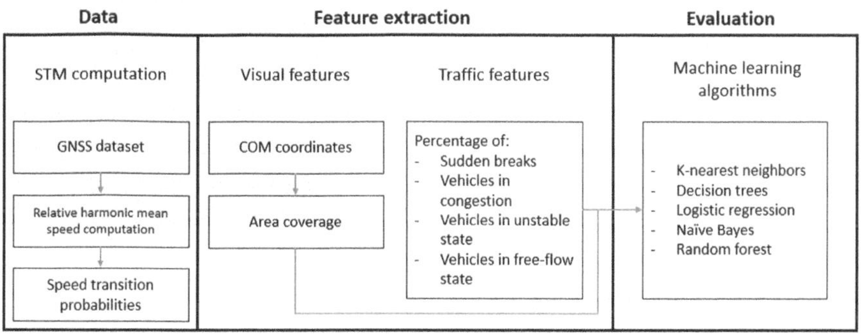

Fig. 1 Overview of the methodology used for STM feature extraction

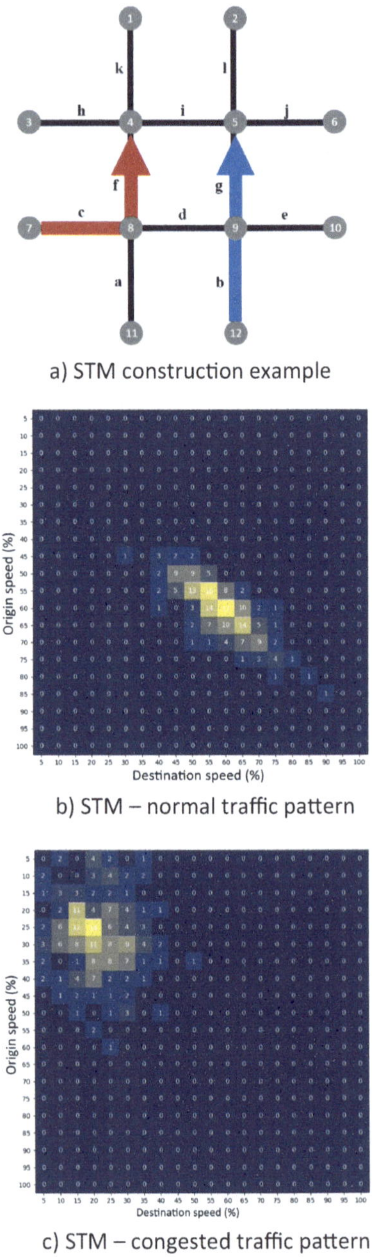

Fig. 2 STM computation and visualization

$$X = \begin{pmatrix} p_{11} & p_{12} \cdots & p_{1n} \\ p_{21} & \ddots & \vdots \\ \vdots & \ddots & \vdots \\ p_{m1} & \cdots \cdots & p_{mn} \end{pmatrix}, \tag{1}$$

where m is the index that represents origin speed values $i = 5, 10, \ldots, m$, n represents the index of destination speed values $j = 5, 10, \ldots, n$, and p_{ij} represents the probability of transition between harmonic speeds i and j on the observed road segment at time interval Δt.

STM dimensions depend on the chosen harmonic speed resolution discretization value. In this paper, 5 km/h is chosen as a discretization value, and 100% is the maximum relative speed value. Then, those values define the STM dimensions of 20 × 20.

2.2 Visual Pattern Feature Extraction

Regarding the previous research [13], the most important visual feature of the STM is the position of the represented traffic parameter. In this paper, we added one more feature responsible for describing the quality of represented traffic patterns. The introduced feature is the percentage of the traffic pattern area coverage compared to the total area of the STM represented in Fig. 3.

The traffic pattern area coverage percentage is a feature that shows the dispersity of the obtained traffic pattern. It can be concluded that patterns with less disparity represent more quality patterns because the position of the pattern can be estimated

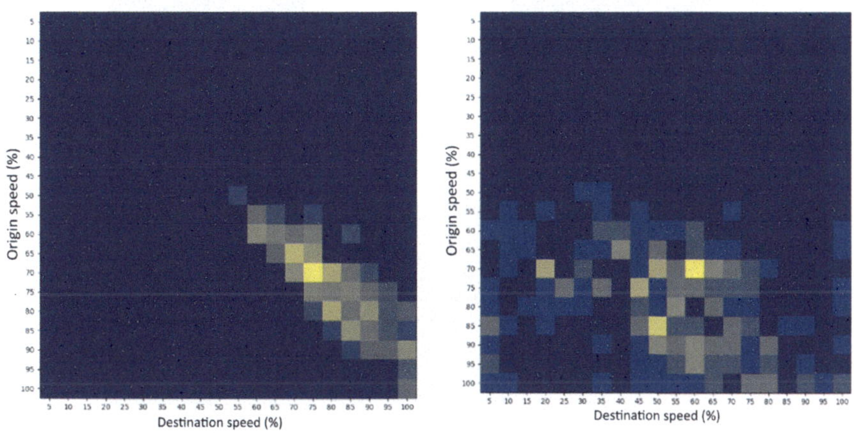

(a) Dense traffic pattern (b) Dispersed traffic pattern

Fig. 3 STM feature: percentage of the traffic pattern area coverage

with higher precision. The patterns with large dispersity represent STMs with almost random behavior. These patterns can indicate more unstable traffic flow or invalid STM if not enough data are provided.

2.3 *Road Traffic Feature Extraction*

Road traffic features are related to the traffic state estimation on the transition represented by the corresponding STM. Figure 4 represents the common areas of the STM used as traffic state estimation features.

Five common STM areas that represent traffic features were proposed: (i) congested area, which represents vehicles that had very low speed on the origin and destination links, (ii) unstable area, that represents vehicles with speeds of 35–70% of the speed limit on the origin and destination links, (iii) free-flow area, that represents vehicles with high speed on the origin and destination links, (iv) sudden breaks area, that represents vehicles that had a large origin and very low destination speed, and (v) intense acceleration area, that represents vehicles with the small origin and very large destination speeds. Other four areas of the STM are not addressed in this research.

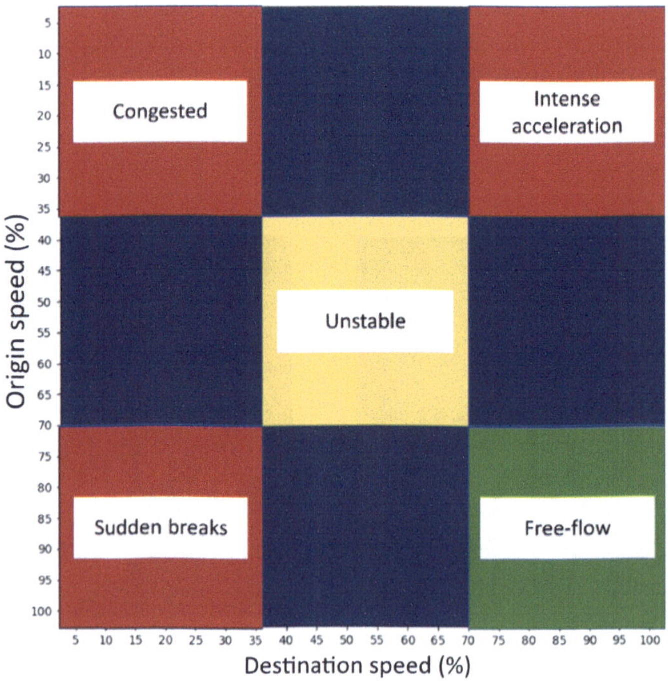

Fig. 4 Common STM areas used as traffic state estimation features

2.4 Data Labeling

The data labeling process is related to assigning class membership to the test dataset containing the STMs using the domain knowledge data extracted from the Highway Capacity Manual (HCM) [5]. HCM defines six levels of service for road segments that are based on vehicle speed values, quantified with letters starting from A to F. LoS A represents free-flow driving conditions with vehicle speeds larger than 80% of the free-flow speed, and LoS F the worst driving conditions with vehicle speeds that are less than 30% of the free-flow speed. The data labeling technique is adopted from Tišljarić et al. [13], where authors labeled STM data into three classes: (i) free-flow, (ii) unstable flow, and (iii) congestion. The labeled data are used as a ground truth when comparing the performance of machine learning algorithms.

2.5 Machine Learning Algorithms

K-Nearest Neighbors (KNN). The KNN is a non-parametric distance-based classification algorithm. It is used for classification or regression tasks with the main purpose of finding the closest centroids to groups of data instances.

KNN can be explained using following steps:

1. Initialize the positions for the K number of neighbors (centroids).
2. Using the Euclidean distance, it is necessary to calculate the distance between all points to K neighbors.
3. For each data point, find the nearest centroid and assign it to the centroid.
4. Move centroids to the point that represents the mean of all data points assigned to the centroid.
5. Repeat 3 and 4 until convergence or a maximum number of iterations is achieved.

Decision Tree (DT). The DT is a supervised machine learning algorithm used for clustering and regression purposes. The goal of the algorithm is to predict a class of a target variable by conducting many simple decision rules that form a tree-like shape. Dataset features are placed in inner nodes of the tree, decision rules are placed in the branches, and each node of the leaf represents the decision result. The algorithm begins with the tree's root node and compares the values of the root attribute with the record attribute. The algorithm based on comparison follows the branch and skips to the next node. The process is repeated until the algorithm reaches the leaf node of the tree.

Logistic Regression (LR). The LR is a machine learning algorithm based on the concept of probability used for classification. It uses a Sigmoid-shaped cost function for the predictions and tends to limit the results in [0, 1] interval representing the probability of the class membership.

Naive Bayes (NB). The NB algorithm bases its classification procedure on Bayes theorem with conjecture about independence among predictors. The Bayes classifier comes down to the assumption that a particular feature in a class is not related to the presence of another feature.

Bayes theorem gives its result from the calculation for the probability $P(c|x)$ from $P(c)$, $P(x)$ and $P(x|c)$:

$$P(c|x) = \frac{P(x|c)P(c)}{P(x)}, \tag{2}$$

$$P(c|X) = P(x_1|c)*P(x_2|c)*\cdots*P(x_n|c)*P(c), \tag{3}$$

where $P(c|x)$ is the posterior probability of class (c, target) given predictor (x, attributes), $P(c)$ is the prior probability of a class, $P(x|c)$ is the likelihood, which is the probability of predictor given class, and $P(x)$ is the prior probability of predictor.

Random Forest (RF). The RF algorithm uses subsets of the obtained dataset to make a series of decision trees and takes their average, consequently resulting in better predictive accuracy. RF does not use a single decision tree but takes predictions from all trees and based on the largest number of prediction votes, determines the result. Thus, the greater the number of trees in the forest, the algorithm has greater precision. The algorithm works in two steps: in the first step, a random forest is created by combining the decision tree, and in the second step, it predicts the outcome for each tree created in the previous step.

3 Results

This section presents an overview of used data and the results of STM feature extraction. Experiments were done using Python programming language, with additional packages: NumPy [4], Pandas [8], and Sklearn [11].

3.1 Feature Extraction

As previously mentioned in Sect. 2, visual- and traffic-related features were extracted from the STM. The feature extraction resulted in a list of features represented in Table 1.

Table 1 List of the extracted STM features

	Feature name	Abbreviation	Usage
Visual features	Center of mass (*x* coordinate)	c_x	To determine the position of the observed traffic pattern represented with STM
	Center of mass (*y* coordinate)	c_y	
	Area	A	To determine dispersity of the traffic pattern
Road traffic features	Percentage of vehicles in congested area	p_{cg}	To determine the percentage of vehicles recorded in each area
	Percentage of vehicles in unstable area	p_{un}	
	Percentage of vehicles in free-flow area	p_{ff}	
	Percentage of vehicles in sudden breaks area	p_{sb}	
	Percentage of vehicles in intense acceleration area	p_{ia}	

3.2 Data

For this research, a real-life dataset was used that consisted of large GNSS data collected by vehicle tracking devices. Data are collected between 2009 and 2014 by approximately 4200 vehicles. The dataset contains 6.55 billion records driven across Croatia. In this research, we used data collected in Zagreb, which is by far the largest city in Croatia, with a population of more than 800,000 citizens. To mitigate traffic seasonality effects, weekend days, the summer months, July, and August are excluded from experiments. Table 2 presents a summary of the collected dataset.

From the obtained dataset, 47,578 valid STMs were extracted, where every STM with at least 100 recorded vehicles is considered valid. After the STM computation,

Table 2 Data summary

Number of GNSS traces	6.55 billion
Sampling rate	100 m/5 min
Time span	August 2008–October 2014
Number of vehicles	4200
Number of road segments (Croatia)	2,000,000
Number of road segments (Zagreb)	86,900
Number of STMs (after filtering)	47,578

Table 3 Example of the extracted STM features formatted as a Pandas DataFrame used as an input to machine learning algorithms

id	stm	c_x	c_y	A	p_sb	p_ia	p_cg	p_un	p_ff	true_cl…
0	0,0,0,0,…	13	15	0.1	0	0	0	0.09	0.51	2
1	0,0,0,0,…	15	10	0.12	0.13	0	0	0.04	0.16	2
2	0,0,0,0,…	15	13	0.08	0	0	0	0.07	0.5	2
3	0,0,0,0,…	13	7	0.17	0.18	0	0	0.09	0.11	2
4	0,0,0,0,…	13	12	0.07	0	0	0	0.09	0.32	2
5	0,0,0,0,…	12	13	0.08	0	0	0	0.11	0.4	2
6	0,0,0,0,…	13	6	0.12	0.3	0	0.01	0.02	0.16	2
7	0,0,0,0,…	14	13	0.1	0	0	0	0.04	0.51	2
8	0,0,0,0,…	13	7	0.12	0.21	0	0	0.07	0.12	2
9	0,0,0,0,…	12	13	0.07	0	0.01	0	0.08	0.38	2

STM features are extracted with the resulting Pandas DataFrame represented in Table 3. For convenience, only the first ten rows are shown.

3.3 *Evaluation*

The results of the STM feature extraction were evaluated on five common machine learning algorithms: KNN, DT, LR, NB, and RF. For each algorithm, the classification results are compared with the ground truth values explained in Sect. 2.3. The goal of the classification was to classify an input STM in one of the three traffic state classes "congestion," "unstable," and "free-flow," based on the ground truth. The input dataset was divided into training and test sets by using the standard ratio of 30% for test and 70% for training.

Figure 5a presents the results of the total accuracy across every considered machine learning algorithm. The best results are obtained by DT and RF with over 0.99 of the accuracy. The worst performing algorithm was NB, with an obtained accuracy score of 0.91. Figure 5b presents the results of the precision, recall, and $F1$ score values, where the precision is computed as $\mathrm{TP}/(\mathrm{TP}+\mathrm{FP})$, recall as $\mathrm{TP}/(\mathrm{TP}+\mathrm{FN})$, and $F1$ as a harmonic mean between precision and recall. Here, TP stands for true positive, FP for true negative, and FN for false negative values. The result confirms DT and RF as the best performing algorithms with precision, recall, and $F1$ scores larger than 0.99.

Figure 6 presents confusion matrices for every considered machine learning algorithm. Most of the algorithms obtained an accuracy score for every class of more than 0.90 (in some cases, visualization framework rounded values larger than 0.9999 to 1). The lowest-performing algorithms were KNN and NB, but only for the class that represented congestion, with an obtained accuracy score of 0.87. These algorithms could not separate classes that represent congested and unstable traffic states due to the similarity between classes.

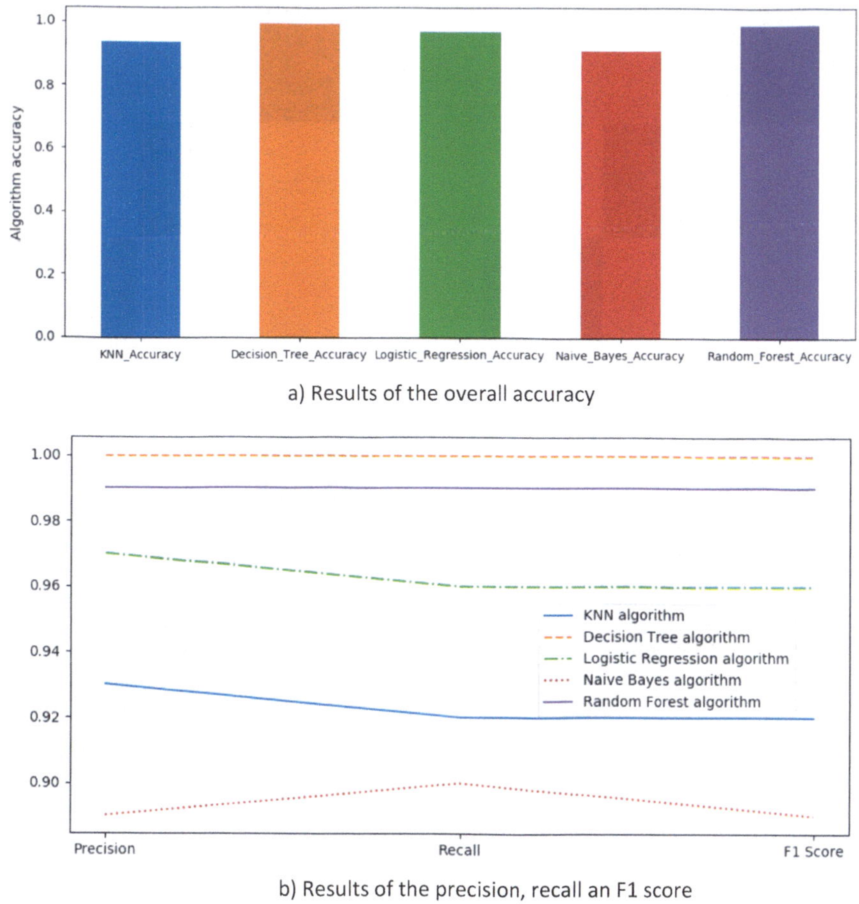

Fig. 5 Comparison of the accuracy for all evaluated algorithms

4 Conclusion

In this paper, a method for feature extraction from STM is presented. Two types of features are presented, visual- and traffic-related. The research goal was to adopt the STM for traffic state estimation using well-known machine learning algorithms. The features are evaluated on the most used machine learning algorithms, with the best results for DT and RF algorithms with the obtained accuracy score of more than 99%. Future research on this topic is related to the automatic extraction of the STM features. The possible method for this task is CNN as it automatically extracts visual features from the given images.

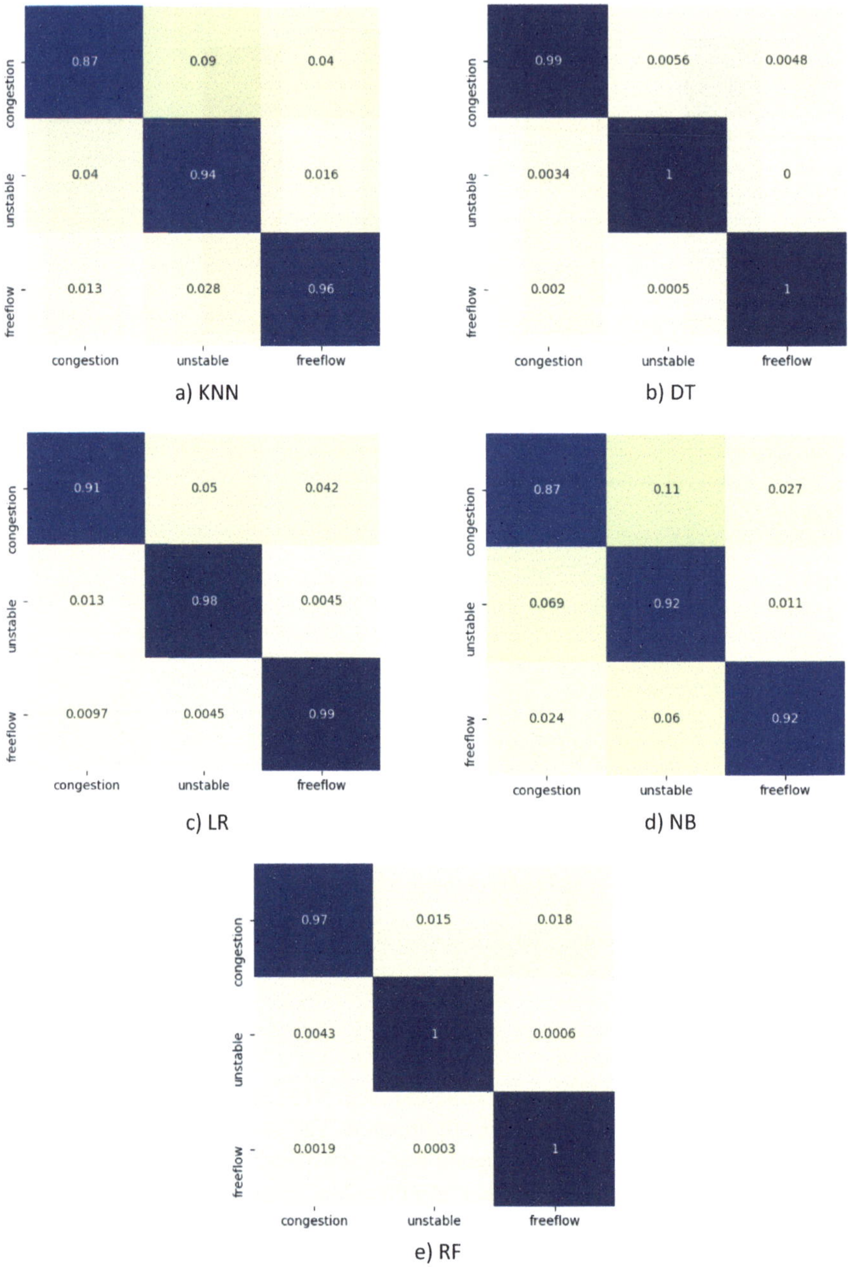

Fig. 6 Confusion matrices for each evaluated machine learning algorithm

Acknowledgements This research has been supported by the University of Zagreb, Student Center, as part of the project "Znanstveno-istraživačke aktivnosti studentske istraživačke skupine SIS-DVA," European Regional Development Fund under the grant KK.01.1.1.01.0009 (DATACROSS), and the University of Zagreb Support Program for scientific and artistic research year 2021 short-term support: "Innovative control strategies for sustainable mobility in smart cities." Data used for this research are collected during project SORDITO, European Regional Development Fund under contract RC.2.2.08-0022.

References

1. Cvetek D, Muštra M, Jelušić N, Tišljarić L (2021) A survey of methods and technologies for congestion estimation based on multisource data fusion. Appl Sci (Switzerland) 11:2306. https://doi.org/10.3390/app11052306
2. Erdelić T, Carić T, Erdelić M et al (2021) Estimating congestion zones and travel time indexes based on the floating car data. Comput Environ Urban Syst 87:101604. https://doi.org/10.1016/j.compenvurbsys.2021.101604
3. Ermagun A, Levinson DM (2019) Development and application of the network weight matrix to predict traffic flow for congested and uncongested conditions. Environ Plan B Urban Anal City Sci 46:1684–1705. https://doi.org/10.1177/2399808318763368
4. Harris CR, Millman KJ, van der Walt SJ et al (2020) Array programming with NumPy. Nature 585:357–362. https://doi.org/10.1038/s41586-020-2649-2
5. HCM (2010) Highway capacity manual, 5th edn. Transportation Research Board, National Research Council, Washington, DC
6. Ma X, Dai Z, He Z et al (2017) Learning traffic as images: a deep convolutional neural network for large-scale transportation network speed prediction. Sensors (Switzerland) 17:1–16. https://doi.org/10.3390/s17040818
7. Mandžuka S, Ivanjko E, Vujić M et al (2016) The use of cooperative ITS in urban traffic management. In: Perallos A, Hernandez-Jayo U, Onieva E, García-Zuazola IJ (eds) Intelligent transport systems: technologies and applications. Wiley, New Jersey, SAD, pp 272–288
8. McKinney W (2010) Data structures for statistical computing in Python. In: van der Walt S, Millman J (eds) Proceedings of the 9th Python in science conference, pp 56–61
9. Nguyen TT, Krishnakumari P, Calvert SC et al (2019) Feature extraction and clustering analysis of highway congestion. Transp Res Part C Emerg Technol 100:238–258. https://doi.org/10.1016/j.trc.2019.01.017
10. Pavlyuk D (2019) Feature selection and extraction in spatiotemporal traffic forecasting: a systematic literature review. Eur Transp Res Rev 11. https://doi.org/10.1186/s12544-019-0345-9
11. Pedregosa F, Varoquaux G, Gramfort A et al (2011) Scikit-learn: machine learning in Python. J Mach Learn Res 12:2825–2830
12. Tišljarić L (2021) Speed transition matrix feature extraction. GitHub repository. https://github.com/tisljaricleo/stm-feature-extraction
13. Tišljarić L, Carić T, Abramović B, Fratrović T (2020a) Traffic state estimation and classification on citywide scale using speed transition matrices. Sustainability 12:7278. https://doi.org/10.3390/su12187278
14. Tišljarić L, Fernandes S, Carić T, Gama J (2020b) Spatiotemporal traffic anomaly detection on urban road network using tensor decomposition method. In: Appice A, Tsoumakas G, Manolopoulos Y, Matwin S (eds) Discovery science. Lecture notes in computer science. Springer, Cham, pp 674–688
15. Tišljarić L, Majstorović Ž, Erdelić T, Carić T (2020c) Measure for traffic anomaly detection on the urban roads using speed transition matrices. In: Proceedings of international convention

on information and communication technology, electronics and microelectronics, Opatija, pp 268–275
16. Vrbanić F, Ivanjko E, Kušić K, Čakija D (2021) Variable speed limit and ramp metering for mixed traffic flows: a review and open questions. Appl Sci (Switzerland) 11:1–26. https://doi.org/10.3390/app11062574
17. Xu L, Wang Y, Yu H, Li H (2015) Feature extraction of urban traffic network data based on locally sensitive discriminant analysis algorithm. In: 15th COTA international conference of transportation, Beijing, pp 2192–2203
18. Yu H, Wu Z, Wang S et al (2017) Spatiotemporal recurrent convolutional networks for traffic prediction in transportation networks. Sensors (Switzerland) 17:1–16. https://doi.org/10.3390/s17071501

Are We Ready for Autonomous Vessels?

Iva Savić

Abstract With the latest amendments to the Croatian Maritime Code, made at the beginning of 2019, "autonomous vessels" made the entrance to the Croatian maritime law. However, the legislator decided to include only the definition of autonomous vessels in the Code, which raises the question of whether the amendment made this way makes any sense. Given the accelerated development of new technologies and the fact that international maritime organizations are working intensively on adapting existing international maritime conventions in the context of the emergence of "autonomous vessels" (of different categories and degrees of autonomy), this paper addresses the phenomena of their inclusion in Croatian legislation. Furthermore, the author analyses the current state of legal regulation of autonomous vessels worldwide and gives recommendations for possible solutions in relation to regulating liability issues in connection to this new kind of vessels in future.

Keywords Autonomous vessels · Unmanned ship · Croatian maritime code

1 Introduction

Although first attempts to define and classify autonomous vessels (according to the degree of autonomy) date back to the end of the twentieth century, with the oldest project dedicated to the research of remote control of ships taking place only in the 1980s in Japan ("highly reliable intelligent ship", 1982–1988) [1], this issue is being discussed in a wider scientific and professional arena since recently [2–4], with the first testing and use of autonomous ships taking place, as well as with the work of international maritime organizations which establish definitions and types of autonomous ships for this purpose (see infra).

The emergence of automation of machines of all kinds is increasingly present in modern industry, and is gaining special attention in the transport industry, especially in relation to autonomous cars and ships (as well as other types of vessels). In order to

I. Savić (✉)
Faculty of Law, University of Zagreb, Trg Republike Hrvatske 14, 10000 Zagreb, Croatia
e-mail: iva.savic@pravo.hr

M. Petrović et al. (eds.), *The Science and Development of Transport—ZIRP 2021*,
https://doi.org/10.1007/978-3-030-97528-9_6

be able to deal with certain issues relevant to legal regulation of autonomous vessels, it is important to consider the current scheme of categories of autonomous ships. There are more classifications of autonomous ships out there (Lloyd's Register, Society of Automotive Engineers (SAE), Verband der Automobilindustrie, NFAS) but we have decided to choose the one set by the International Maritime Organization (IMO), since it has been designed inside of an organization which enjoys high level of credibility in a global arena when it comes to maritime law issues. The IMO distinguishes 4° of autonomy that has been developed for the purpose of conducting an analysis of existing maritime international conventions in the context of the regulation of autonomous ships [5]. According to this classification, we can distinguish:

- ships with some automated operations, with a crew (located on board) that manages and controls deck systems and functions (so-called first degree),
- remotely controlled ships, operated and controlled by shore-based operators, which may be: a) with a crew on board (so-called second degree), or b) without a crew on board (so-called third degree),
- fully autonomous ships, with a built-in operating system that can autonomously make decisions and take actions (taking into account the possible consequences and risks of thee) (so-called fourth degree).

It is important to note that one ship may have different degrees of autonomy during one operation, but it is necessary that it has a certain degree of autonomy (any of the four listed) during the entire voyage. It is also important to note that although the IMO's classification includes ships which might have a crew on board, the shipping industry, but even more the insurance industry, are pointing to differences between unmanned and autonomous vessels and therefore those cannot be used as synonyms.

However, IMO's classification actually reflects the current state of the industry pretty well. Although it is still in its infancy, this industry is growing very fast, and is working intensively on the development of new technologies. In addition to the renowned leaders of this revolution—Kongsberg, Rolls Royce and Japanese shipbuilders—many other companies are coming forward in recent years developing certain components important for the implementation of autonomous ship projects (such as Shone, Yara). Thereby maritime industry is experiencing another similar trend of smart ships that however differs from autonomous vessels [6]. What distinguishes these two phenomena is a "philosophy", both technological and business one that is behind each of them. Namely, while the so-called smart ships are characterized by (only) a higher level of digitalization, autonomous ships are characterized by a completely different system of operation, based on the so-called machine learning, which represents a type of applied artificial intelligence [7]. Precisely because of this important difference, autonomous ships need a special approach in regulation.

2 Use of Autonomous Vessels in Transport of Goods and Passengers by Sea

Norway is at the forefront of the development of autonomous vessels in the world, with a long tradition and high stimulus in finding innovative solutions in this sector. Thus, a geographical area (fjord Trondheim) has been designated in Norway where autonomous ship testing is carried out by the Norwegian Maritime Administration, the Norwegian Coast Guard, as well as various participants from industry and research centres and universities [6]. However, interest in the autonomous mode of maritime transport is growing in other European countries as well. The first EU project (MUNIN Vision 2012–2015) practically consisted of an extensive study on autonomous merchant ships, whose task was to analyse the possibility of transforming Handymax, a ship intended for the transport of dry bulk cargo—into an unmanned ship. Since then, the number of new research, studies and projects has been constantly increasing, and during last two years, first autonomous ships have been demonstrated within several projects.

On the new EU road to autonomous ships, apart from Denmark and Finland, which along with Norway, are at the forefront of the autonomous maritime "revolution" in Europe, private entities from Belgium, Italy, France, Scotland (and Norway) have formed a consortium goal set within the EU project called AUTOSHIP Autonomous Shipping Initiative for European Waters [8]. The main goal of the project is to accelerate the transition phase towards the next generation of autonomous ships in the EU. The project plans to build and commission two different autonomous vessels, which goes in the direction of the EU's maritime policy (that focuses on reducing road congestion and related pollution), and since 2001 has been developing the concept of a "sea highway". This concept has been introduced in White paper, 2001, after which it was developed throughout several other documents, inside the TEN-T policies [9]. On the same track is the latest EU project AEGIS (Advanced, Efficient and Green Intermodal Systems) [10], aiming to create a new "water" transport system in the EU in the following years, in order to connect autonomous ships and flexible systems used in inland waterway transport with ports and their automated cargo management services and other digital technologies [11].

But in parallel with the projects mentioned, autonomous ships have become a reality in the EU in the last few years. In December 2018, Finnish carrier Finferries and Rolls-Royce demonstrated the result of their multi-year project, the fully autonomous ferry Falco, the first of its kind in the world. The ferry ride was demonstrated in the Finnish archipelago south of the city of Turku, which transported 80 VIP guests between the cities of Parainen and Nauvoo. The transport itself was performed with autonomous remote control without the participation of the crew (fully autonomous control), and during the functioning of special sensors was demonstrated showing that Falco can detect objects alone, which makes him able to avoid collisions and perform automatic docking and berthing [12].

In 2020, Kongsberg Maritime performed its first custom ferry service during its regular ferry service. At a short distance between the two docks, the Bastø Fosen

VI ferry carried passengers and vehicles, and the transport was carried out with full automatic control. A trial phase for the Bastø Fosen VI lasted for six-month, during which for most services the ship was controlled by an automatic system; however, captain and crew were retained on the ship's bridge, and it is planned for them to stay even after reaching higher level of the ship's autonomy. The reason for this is that the ship is not fully automated. As it is stated, if another vessel or object on the road is spotted, an alarm will go off, after which the captain of the ship will take control of the situation. This year, the installation and testing of a collision prevention system (consisting of radar and electro-optical sensors) is planned [13].

In addition to this, Kongsberg is also working on the first autonomous container ship (which will also be entirely electrically powered and will not generate any emissions). They are responsible for development and delivery of all key enabling technologies including the sensors and integration required for remote and autonomous ship operations, in addition to the electric drive, battery and propulsion control systems. According to the announcements, YARA Birkeland will move to full autonomous operation by 2022 [14].

Although the list of autonomous ships (projects) does not end here, in addition to ferries, container ships and other cargo ships, autonomous operation programs are being developed also on other types of vessels as well as underwater facilities (e.g. waterjets with the smart operation system that includes "CanMan Touch" joystick). The same can be said for autonomous vessels on inland waterways. Several projects which have the intent to introduce autonomous technology in inland waterways shipping are underway, and regulatory work on adapting the current legislative framework is taking place both at the national and international level (UNECE).

All of this points to the fact that technology is evolving fast and that remotely controlled ships are to become a reality in the near future. Svitzer expects that more vessels will be controlled from the shore than from the ship in the next five to ten years [15]. Given these trends we can say it is already (too) late with the adoption of legislation on this new model of transportation. However, the work on this has slowly started, and again we can point to the Norwegians whose Maritime Authority has begun to adapt its national legislation that could serve as a legal basis for operations of automated seagoing vessels.

3 The Work of International Organizations on the Preparation of National Legislation and International Regulations

Entire initiative in the development of autonomous vessels comes from technical professions that have advanced work on the development of smart systems, artificial intelligence, special kind of sensors, and various other related tools needed for autonomous ship operations. But for smooth inclusion in the maritime transport autonomous ships and their users need a legislative framework. Given that historically maritime industry and transport have been marked by people, and that we automatically link people—captains and crew—to the ship (and other seagoing vessels), it is not odd to have the impression that existing maritime law, both international and domestic, is not acceptable for the new types of vessels and maritime transport. Such an instinctive reaction by experts and scientists around the world has logically resulted in taking the first steps towards adapting the existing international maritime law regime. This is led by the International Maritime Organization (IMO) and the International Maritime Committee (CMI), whose activities will be presented and studied in this chapter in more detail.

3.1 The Work of IMO

Primary interest of the IMO relates to the issues of safety of autonomous ships and their operations, and to the existence and role of the crew on such vessels. In this regard, the Maritime Safety Committee (MSC) and the Legal Committee (LEG) have taken on certain tasks.

From 2018 to 2020 MSC has led a large project on autonomous ships. It is now being waited for the MSC to address the issues of compliance of current (IMO) regulations with the new type of ship, primarily from the aspect of safety and environmental protection. Related issues of liability for damage caused by the operation of such ships would be in the domain of the Legal Committee (LEG). In order to carry out this great tasks, the IMO has defined the "Maritime Autonomous Surface Ship" (MASS) as a ship "which, to a varying degree, can operate independently of human interaction" [16]. Based on the "different degrees" of possible autonomy, MSC has further established the classification of autonomous ships (which has already been discussed supra).

The first step was to determine the applicability of the provisions of the IMO instruments to autonomous ships and their impact on possible autonomous ship operations, in the context of the following areas or issues, and the international instruments governing them: security and protection (SOLAS); collision (COLREG); loading and stability (Load Lines Convention); training of seafarers and fishermen (STCW Convention and STCW-F Convention); search and rescue at sea (SAR); tonnage convention; container safety (Convention on Safe Containers); and conventions governing passenger ships for special types of passenger transport by sea (SPACE STP, STP). That was followed by the analysis and regulatory scoping exercise with the aim to define the best way to regulate autonomous ship operations under the existing and/or new regulations.

As early as 2019, the MSC adopted transitional guidelines for conducting test operations of autonomous ships [17], which require tests to be performed in a way that will ensure at least the same level of safety, security and environmental protection as required by the relevant instruments. Furthermore, guidelines require tests to properly identify possible risks and envisage measures to mitigate them. The guidelines further prescribe certain conditions on the part of the operator, those on board the autonomous ship or outside it, as well as taking appropriate steps in the form of protection against possible cyber-attacks. In parallel, the MSC has continued with the regulatory scoping exercise and although the COVID-19 pandemic-related issues have delayed the work, the outcome of this exercise was finally discussed and completed by the MASS Working Group in May 2021 (MSC 103). Results of this complex analysis have shown that there is a lot of work ahead for IMO in addressing the MASS regulation, on so many levels. Next we will address only the highpoints.

First, it has been concluded that some high-priority issue will first have to be addressed at a policy level in order to help regulatory bodies define and shape the future work on MASS regulatory framework. Those include clarifying the meaning of terms "master", "crew" and "responsible person"; addressing functional, operational and monitoring requirements of the remote control station/centre; addressing qualifications, responsibility and the role of remote operator as seafarer; and working on terminology issues (MASS glossary).

Other issues that have also been recognized as important to solve can be considered more as "technical". They include different common potential gaps and/or themes that appear in one or more maritime law conventions. Those are provisions containing manual operations and alarms on the bridge; provisions related to actions by personnel (firefighting, cargoes stowage and securing and maintenance); certificates and manuals on board; cybersecurity; watchkeeping; implications for search and rescue; and information required to be on board for safe operation.

Finally, after finishing a complex analysis of the provisions of all above listed maritime conventions in the light of MASS (operations), it has been noted that "the most appropriate ways of addressing MASS operations" could preferably be through the development of a MASS instrument, which could take the form of a "MASS Code", instead of amending individual instruments. That Code would cover the goals, functional requirements and corresponding regulations for all four degrees of MASS, following a goal-based approach, in line with the IMO's Guidelines [18]. It is interesting to note that drafters have recognized it necessary to make an agreement on the use of terminology and is a policy decision. One of the issues to be addressed was considered to be the re-evaluation of the degrees of autonomy, taking into account the lessons learned during the RSE. This work could include the development of a glossary.

Other than that, one of the most important issues that the LEG, other IMO's committee, is dealing with is the one of crew or seafarers whose role, with the advent of autonomous ships, changes significantly. In the past two years, the LEG has started working and determined which conventions will be analysed from the aspect of autonomous ships, the form according to which data and documentation will be collected, and the plan and program for the implementation of this work. The aim of this analysis is to determine the extent to which the regulation of autonomous ships will affect the existing maritime law regime, and in this regard to determine (1) whether or not it applies to autonomous ships for each of the provisions of the selected conventions, (2) if applicable, whether it disables (prohibits) operations by autonomous vessels, or (3) if it does not prohibit autonomous vessels, whether changes, clarifications, etc. are necessary. This analysis should include autonomous ships of all degrees of autonomy. Following the completion of this process, the LEG would determine in the next step the most appropriate way to address the issue of autonomous ship operations, depending on the results of the previous analysis.

3.2 *The Work of CMI*

CMI decided to make a detailed analysis of national regulation in its state parties before taking the next step in creating new regulations. In the focus of the analysis was the positioning of autonomous ships in the context of the existing national legislation of the state parties, and their interpretation of maritime international law (conventions) from the perspective of autonomous vessels.

Nineteen countries (national maritime law associations) have made their responses to the CMI questionnaire publicly available, including Croatia [19]. The questionnaire prepared by the CMI contains several key questions related to the application of existing maritime law regulations to autonomous ships, and this paper will assess the most important ones, followed by the analysis of the differences in the approach of individual national legislations.

First, and somewhat the most important question regards the qualification of an autonomous ship (in excess of 500 GT) in national law of CMI members states, i.e. their merchant shipping law, in terms of "recognizing" the status of that kind of cargo ship as an autonomous ship, if it is without a master or crew on board, and either (a) controlled remotely by radio communication, or (b) controlled autonomously (by, inter alia, a computerized collision avoidance system, without any human supervision).

Interestingly, the vast majority of countries gave an unequivocal positive answer to this question (Belgium, Italy, Japan, France, Greece, Malta, the Netherlands, Panama, USA, Singapore, Great Britain and Venezuela). Other countries gave a positive interpretation of their own regulations (for example Argentina, Canada and Spain), while Croatia gave an explicitly negative answer, referring to the Art. 76, para. 2 of the CMC.19. This provision stipulates that in order to determine a ship as seaworthy, it is necessary that she has the prescribed number of crew members with appropriate certificates of competence on board. If the responders had strictly adhered to the definition of a ship as found in Art. 5 para. 4 CMC, they could have followed the line of the majority of other states and bring the autonomous ship as defined in this questionnaire under the existing definition of a ship as found in the CMC. The cited provision states as follows: "A ship, other than a warship, is a vessel intended for navigation at sea with a length of more than 12 m and a gross tonnage of more than 15, or authorized to carry more than 12 passengers(…)" [20]. The ship's ability to sail (seaworthiness), which was in their focus, was not realistically at stake here. It is interesting in this context to consider the letter of prof. Sozer [21], sent to the CMI on the occasion of this questionnaire, with regard to definition of the ship (vessel). In this letter he essentially reminds us that shape and usage of ships has changed over time, and that followed by technological changes, innovations and issues that ships caused to other subjects and the environment, it is hard to find a comprehensive definition for it. Although in his letter prof. Sozer compiled a complete list of conventions that contain or do not contain a definition of a ship and made his comments on the definition of a ship in the existing international maritime conventions, making this analysis thoroughly, we should carry one important thought from his letter—when thinking about future autonomous ship regulation, it is important to distinguish essential features of the various types of services for which ships are used today.

Another thing that should be addressed in the context of Croatia's response to the questionnaire is the provision of Art. 148 para. 3 CMC in which it is written that the master "is obliged to be on board during the voyage". Given this, a ship whose master is outside the ship—and which is one important feature of autonomous ships—and manages it remotely, in Croatian maritime law would not enter the definition of a "ship" according to the CMC. It is interesting to compare the situation and the response given by Brazil, since they have very similar laws. Namely, Brazil considered decisive the fact that Brazilian law defines a ship as a thing—without the additional condition of the existence of the commander and crew on the ship. Therefore, in their opinion, autonomous ship can fall under the existing definition of a ship. However, Brazilian law also specifically prescribes a requirement for the presence of the master and crew on board, in order to navigate and transport passengers and their luggage, but this did not make them change the answer given to this question [22]. Given all of the things mentioned, we can conclude that this may have been a motive for the Croatian legislator to introduce a special definition for an autonomous vessel in the CMC, separately from the main one.

Following this, the second important question of the questionnaire is related to the issue of the control of the ship in the context of the crew, and seeks an answer to the question of whether different categories of possibly responsible persons can be validly considered as a ship master in accordance with national regulations. CMI, in particular, assumed 3 types of possible masters of autonomous ships: (1) the master who manages the ship remotely, from the shore (chief onshore remote-controller), (2) the programmer of the autonomous ship (Chief pre-programmer of an autonomous ship) and (3) the person envisaged as the master, but only on the paper (responsible person), which is not directly involved in operations on board ship. The very fact that as many as three, very different, concepts were assumed, resulted in very diverse responses and useful comments from the countries that participated in the analysis.

In the first category, which deviates the least from the common notion of a master (since it presupposes his existence and work, but from a distance), it is already clear that most national regulations included require the physical presence of the master on board, or set some additional requirements that indicate the need for his physical presence during maritime venture. Nevertheless, several states have answered that under their law, a person who is not physically on a ship but manages it from the shore (remotely), might be considered as the master. Such an answer was given by Panama and the United States, but besides them, several others (Belgium, France, Great Britain and Australia) have left that possibility open. Singapore's position, although not clear, might be taken close to this. Namely, based on the interpretation of the provision of Art. 2 of the Maritime and Port Authority of Singapore Act, in their opinion it is not clear whether any of these categories can be classified as masters. However, after looking at the cited provision (which only requires the master to "have a charge or command"), it can be assumed that the first category could possibly be classified as a master, however depending on the interpretation of words "having charge or command". Namely, France, which has a very similar condition in its law (according to them, the person of the master must in fact be in command of the ship), said it "is not inconceivable" that the chief onshore remote-controller could be considered as a captain. The Australian national law has a similar wording, and Australians are of an opinion that the word "command" refers to the control of the vessel, its navigation and operations, and that it does not in itself require the presence of the master on deck. Therefore, if that remote control was sophisticated enough that the person remotely operating the ship had similar control over the ship as if he were physically on board, in that case he could be considered as the master (of an autonomous ship).

With regard to the second and third categories, no state has explicitly stated that such types of roles could be considered as maters. Only a little space for this possibility, given the second category (chief pre-programmer of the autonomous ship), was left by Belgium, but only to the extent that it is not explained in questionnaire what exactly is considered under that phrase. The Belgians concluded that if the job of a pre-programmer is limited to inputting data entry before the start of the voyage, it would be hard to consider him having control over ship (later) and being the master.

To conclude: 12 out of the 19 states (63%) eliminated the possibility that any of the three categories listed could fall under the definition of a ship's master according to their applicable national laws. These are: Croatia, Greece, Italy, Malta, Spain, the Netherlands, Brazil, Japan, South Africa, Canada, Argentina and Venezuela. The most common explanation behind this reasoning is that regulation requires physical presence of the master on board, either in the definition, or through some other provisions governing his duties. Just to mention as the example, the response regarding the Croatian CMC included the provision of Art. 125, para. 1, which stipulates that "the ship's crew consists of the master and other persons embarked to perform duties on the ship and entered in the crew list". However, to this, we could add Art. 148 para. 3 which prescribes that the master "is obliged to be on board during the voyage". Of the remaining seven states, only two states (10% of total) believe that a person who remotely controls a ship can be considered as a master, and the remaining five (26% of total) in some way leave this possibility open. However, the most valuable and interesting was to observe the diversity of legal arrangements and assumptions that national laws set as a condition on the part of the master, as well as their different interpretations of sometimes very similar conditions. This clearly points to potential problems in taking further regulatory steps in this area, but at the same time proves there is a need to regulate the smallest details related to these, for maritime law, new concepts.

Although the position of a crew member may at first seem similar to that of a ship's master, which is the case in many legislations (including Croatia, Greece, Italy, Malta, Spain, South Africa, Japan, Canada and Venezuela), as both roles require the physical presence on board, the questionnaire responses show that states do not always equal conditions for both role of a crew member and the one of a master. Singapore, which was vague in answering the questions regarding the master, made it clear that persons who remotely control the ship could not be considered to be crew members. Interestingly, Panama, which was open to new concepts in terms of the master's role, also left open the possibility for crew members to be considered persons remotely operating the ship, but conditioned this by the type of work assigned to them. Argentina and Netherlands gave a very similar answer to this one, too. France, in contrast to the interpretation given in relation to the master, explicitly stated here the possibility of a person operating from a distance being considered a crew member. The interpretation given by Belgium is particularly valuable, as it went a step further from a strict interpretation of commercial maritime law (which does not explicitly require the physical presence of a crew member on board) and referred to the provision of another national regulation that asks for the specific capabilities of a crew member (Circular, 1.8.2014). Finally, Australia, United Kingdom and the United States, in contrast to the position taken with respect to the person of the master of the ship, explicitly replied here that such a possibility did not exist for a crew member.

4 Autonomous Vessels in CMC

In accordance with the provisions of Art. 5 para. 1 CMC, an autonomous vessel is defined as "a vessel which, depending on the degree of automation and the requirements for direct supervision of the permanent service, can sail without a crew on board or with a reduced number of crew members". From this definition of autonomous vessels, we can conclude that according to the Amendments to the CMC, autonomous vessels are defined by 3 properties:

1. automation,
2. (non) existence of the need for direct supervision of the permanent service,
3. the possibility of sailing without a crew, or with a reduced number of crew members (embarked on the vessel).

A closer look at the definition tells us that not all of the three elements listed need to exist for a ship to be autonomous, but the third is actually a characteristic of an autonomous ship that appears as a consequence of fulfilling other two conditions. Furthermore, based on the wording of this provision, it could follow that a ship with a certain degree of automation and sailing with a crew could also enter the definition of an autonomous ship (CMC prescribes only the "possibility" of a ship to sail without a crew). In addition, it is not clear from this whether there must be "direct supervision of the permanent service" in any case, or perhaps, given the "degree of automation", this is not necessary!? This also makes it unclear in which way should "direct supervision of a permanent service" exist—should that service include humans or some software (program) can take over the job of supervision; Is the criterion immediately met even if the service located on shore performs remote surveillance over the ship via sophisticated telecommunications channels? These elements are not defined precisely, and since they allow different levels (degrees) of autonomy, it is not clear from the provision what would a "lower" and what would an "upper" limit of autonomy CMC set. Given the rapid development of technology in this area and the lack of experience in performance of this type of ship, the broad definition in itself is not bad overall. But a "cleaner", clearer definition, which descriptively defines an autonomous vessel would be better in this situation. Finally, taking into account all previously stated, the question arises whether it was necessary to include this definition of autonomous vessel in the CMC, especially since the application of the provisions of the CMC to this category of vessels is not specified.

In this regard, it should be noted that there were some changes made to the provision of Art. 2 CMC, which defines Code's scope of application. Namely, in accordance with the new Art. 2 CMC, unless otherwise provided by the Code, its provisions regarding ships shall also apply to other maritime objects (except warships). Although an autonomous vessel could fall under the definition of a vessel (a vessel is a maritime object intended for navigation at sea), just as an autonomous ship could fall under the existing definition of a ship (more supra, Chap. 3), it is not entirely clear why the legislator opted for this solution. Namely, according to the CMC a vessel can be a "ship, warship, submarine, yacht or boat". Putting in connection Art. 5 item 27 of the CMC and item 1, we can conclude that according to the current regulatory situation, an autonomous vessel cannot be considered a vessel (in the manner regulated by the CMC), although its name would suggest that it is precisely a vessel (certain types). This also has the consequence that under the current version of the CMC, provisions regulating maritime objects will not apply to autonomous vessels. This means that although Croatian maritime law has recently recognized the category of autonomous vessels, it does not contain provisions that would further on regulate these facilities in any aspect.

However, regardless of the way in which autonomous ships (vessels) are arranged in the CMC, it is necessary to consider whether, given the (non) existence and (non) use of such vessels in our area, there is any need or sense to have them regulated in Croatia. Namely, as stated supra in Chap. 2 of this paper, the leading international maritime organizations have not yet found a way to adequately regulate the framework for autonomous ships in the context of existing international instruments. Likewise, with the exception of Norway and Denmark, EU Member States are not yet addressing this issue. The primary reason for this is insufficient knowledge of how autonomous ships work, both due to the new type of technologies they use, and due to the fact that most of them are still used exceptionally and / or are in the testing phase. However, what could already be interesting in Croatia on the topic of autonomous ships is their business model, primarily when it comes to passenger transport, and it would therefore be advisable to devote attention, knowledge and money in that direction.

5 Liability for Damage Caused by the Use of Autonomous Ships

Perhaps the most important and controversial issue related to the use of autonomous ships that should be considered is the issue of liability for damage caused by an autonomous ship. Given the scope of this paper, not all types of possible damages and liable persons will be analysed in detail, but the issue of liability will be considered in principle, given the current legal regime in the light of insertion of autonomous vessels in Croatian maritime law, and existing research on these issues. Given the lack of international regulation that explicitly regulates the issue of liability of autonomous ships, in this analysis we will also take into account selected solutions and concepts developed in the car industry. However, these should be used carefully, taking into account the setting and specifics of maritime transport and the fact that there is a generally accepted differentiation between several types (degrees) of autonomy for both cars and autonomous ships.

We encounter various types of liability regimes in maritime transport, each of them for a different type of damage caused by ship operations: to passengers (and their luggage), to shippers (their cargo), to environment and to third parties. All of them can be, and often are, victims of damage caused by the usage of a ship or by a ship itself. In all cases mentioned there is a legal framework in place at domestic and international levels. A general rule is that the responsibility is placed either on the carrier and/or on the shipowner. This being addressed from the perspective of an autonomous ship causing any such damage raises the question as to whom or "what" liability should be attached, and whose actions should be observed in the process of determining responsibility, and according to which rules.

Here, we will take a look at possible scenarios with reference to existing solutions regarding liability for damaged caused by autonomous cars. Namely, seemingly faster progress of autonomous vehicles and the planned commercialization of their use in the US and in SE Asia (especially China, Japan and South Korea) have prompted earlier regulations. Nevertheless, Germany has come to the forefront with introducing the new law on autonomous driving (Gesetz zum autonomen Fahren), which sets the legal framework for integrating autonomous vehicles (level 4) in regular traffic in Germany in 2022. Similarly, in the United States there have been several regulations related to the development of autonomous technologies in the automotive industry and autonomous vehicle operations adopted in the last few years. In the context of the manufacturer's responsibility for the defective product, the comparison in the position of the manufacturer of both types of vehicles (cars and ships) is unquestionable. In case of both types of vehicles, answer on the question of a different liable person—be it a car driver or ship master/operator—will depend on the level of autonomy of a vehicle in question and parallels could be drawn only between the same levels of autonomy. In addition, it is important to take into account the specifics related to the different degrees of autonomy of ships and their difference according to the defined autonomy (levels) of cars as well as the environment in which maritime transport takes place. The Association of Automotive Engineers distinguishes several degrees of autonomy, depending on the driver's involvement in the driving process. After level 0, which in reality means the absence of any vehicle autonomy since the driver manages all driving tasks himself, there are 5 more levels. Only at the 3rd level of automation, where the need for regular driver control over the steering wheel ceases, and he is expected to be ready to take control when he is notified, we are talking about an autonomous vehicle [23]. However, this still high level of driver involvement in the driving process can be problematic from the aspect of vehicle autonomy, which is why manufacturers are turning to the production of autonomous vehicles of the 4th and 5th degree of autonomy. The case is similar when it comes to MASS—the issue of determining a person liable for the occurred damage becomes complicated only at the third and fourth level of autonomy. Therefore, all of those things said about autonomous cars could be taken into account also when discussing liability in the context of MASS.

When talking about the contractual liability for damage caused by an autonomous ship, and in the context of the established regulation on carrier's liability for damage to things carried on board ship and damage caused to passengers, we meet at least two potentially responsible persons: a) the carrier, which will now be represented by a different "kind" of a ship master, probably a person who remotely controls the ship (if it is not a fully autonomous ship) [24, 25] the manufacturer of the autonomous ship, and/or manufacturers of certain parts of the MASS, including hardware and software developers, in the context of liability for defective product. In that sense it will be important to determine who is considered to be the manufacturer in relation to the injured party, since his right to compensation will depend on this. Here we should also take into account the Council Directive 85/374/EEC relating to liability for defective products and solution found in selected national laws regulating this issue [25, 26]. It is important to mention that there are several different individuals that participate or may participate in the production of different parts of an autonomous ship—software manufacturer, programmer, manufacturer(s) of various sensors and of GPS, etc., as well as that there is also an approach to a fully autonomous ship as being its own master [24].

Besides those mentioned, in the context of the second- and third-degree autonomy, onshore, remote controllers (MASS personnel) should also be taken into account as possibly responsible for damage caused by (operation of) the MASS while taking the role of the master of the ship. Hence, their role and responsibility should be defined precisely in future. This has also been recognized by the IMO in its Outcome of the Regulatory Scoping Exercise for The Use of Maritime Autonomous Surface Ships (2021). IMO has recognized that defining a master and (onshore) "responsible person" is considered to be a common theme identified in several maritime law instruments as a potential gap which should be solved [18].

Regarding non-contractual liability for damage resulting from the use of an autonomous ship, and in the context of the established non-contractual liability regime where the shipowner is liable for damage to third parties, there will be no dilemma in the application of the provisions on shipowner's liability. What can be corrected is the very basis of liability, i.e. the principle according to which the shipowner will be liable for damage (strictly or on the basis of presumed liability).

However, if the regime is to be adjusted to the specificities of autonomous ships and their use—which would be to be expected—other solutions should be considered as well.

6 Conclusion

"Autonomous ships are not conventional unmanned ships" [6] and their appearance in the maritime industry represents in the fact the emergence of a new business model of shipping. These two things, although seemingly very clear, are crucial for understanding the issue of autonomous vessels and, furthermore, their legislative regulation.

In addition to the emergence of new technologies used for the construction, launch and operation of (autonomous) ships, completely new business models in maritime transport are associated. In such, completely new circumstances, with completely new characteristics of transport, new types of risks appear. Autonomous ships conceptually eliminate the problem of human error (as one of the main causes of collisions and other maritime accidents), but therefore carry other types of risks, such as technical (systemic) errors, cyber risks, and other types of risks that are attributed to the new types of ships.

The work of international maritime organizations and legal experts and scientists in the field that began only a few years ago shows that we are at the very beginning of finding answers to the questions of defining autonomous ships, identifying persons responsible for damage caused by the use of autonomous ships, and insurance. I believe that, given this, and all the knowledge so far, and without the objective need for legal regulation of autonomous ships in Croatian law at this time, the legislator could wait a little longer with this amendment to the CMC. However, given that autonomous vessels have already been given a place in the Maritime Code, this bold decision should have been accompanied by additional amendments that would regulate other related issues, and ultimately those related to liability for damage and insurance. However, obviously it was still too early for such a scenario.

References

1. Norwegian Forum for Autonomous Ships (NFAS), at http://nfas.autonomous-ship.org/why-en.html. Last accessed 25 Apr 2021
2. Ringbom H (2019) Regulating autonomous ships—concepts, challenges and precedents. Ocean Dev Int'l L 50:141–169
3. Klein N, Guilfoyle D, Karim MS, McLaughlin R (2020) Maritime autonomous vehicles: new frontiers in the law of the sea. Intl Comp Law Q 69:719
4. Guerra S (2017) Ready about, here comes AI: potential maritime law challenges for autonomous shipping. 30 U.S.F. MAR. L.J. 69

5. Regulatory Scoping Exercise for the Use of Maritime Autonomous Surface Ships (MASS) (2021) Report of the Working Group, MSC 100/WP.8, 6 December 2018. https://www.setfords.co.uk/wp-content/uploads/MSC9955.pdf. Last accessed 05 May 2021
6. Transport 2040: Autonomous Ships: A New Paradigm for Norwegian Shipping—Technology and Transformation (2021). https://commons.wmu.se/cgi/viewcontent.cgi?article=1072&context=lib_reports. Last accessed 27 Apr 2021
7. Wu J (2021) AI, machine learning, deep learning explained simply. https://towardsdatascience.com/ai-machine-learning-deep-learning-explained-simply-7b553da5b960. Last accessed 30 Apr 2021
8. AUTOSHIP (2021) https://www.autoship-project.eu/. Last accessed 25 Apr 2021
9. Guidelines TEN-T (2013) (Regulation (EU) No 1315/2013, OJ L 348, p. 1)
10. AEGIS (2021) http://aegis.autonomous-ship.org/. Last accessed 5 May 2021
11. Aegis Project Combining the Potential of Autonomous Ships with Automated Terminals Kicks Off. https://shipip.com/aegis-project-combining-the-potential-of-autonomous-ships-with-automated-terminals-kicks-off/. Last accessed 5 May 2021
12. Rolls-Royce and Finferries Demonstrate World's First Fully Autonomous Ferry (2021) https://www.rolls-royce.com/media/press-releases/2018/03-12-2018-rr-and-finferries-demonstrate-worlds-first-fully-autonomous-ferry.aspx. Last accessed 5 May 2021
13. Automatic Ferry Enters Regular Service Following World-First Crossing with Passengers Onboard. https://www.kongsberg.com/maritime/about-us/news-and-media/news-archive/2020/first-adaptive-transit-on-bastofosen-vi/. Last accessed 25 Apr 2021
14. Autonomous Ship Project, Key Facts About Yara Birkeland (2021) https://www.kongsberg.com/maritime/support/themes/autonomous-ship-project-key-facts-about-yara-birkeland/. Last accessed 10 May 2021
15. Svitzer Expects to Control More Ships from Land in 5 to 10 Years (2021) https://shippingwatch.com/carriers/article11749059.ece. Last accessed 25 Apr 2021
16. IMO Takes First Steps to Address Autonomous Ships (2021) Briefing: 08 25/05/2018. http://www.imo.org/en/MediaCentre/PressBriefings/Pages/08-MSC-99-MASS-scoping.aspx. Last accessed 27 Apr 2021
17. Interim Guidelines for Maritime Autonomous Surface Ships (MASS) Trials
18. IMO (2021) Outcome of the regulatory scoping exercise for the use of maritime autonomous surface ships (Mass), MSC.1/Circ.1638, 3 June 2021. https://www.imo.org/en/MediaCentre/PressBriefings/pages/MASSRSE2021.aspx. Last accessed 20 June 2021
19. Croatian Maritime Law Association (Hrvatsko društvo za pomorsko pravo) (2021) CMI questionnaire on unmanned ships. https://comitemaritime.org/work/mass/. Last accessed 5 May 2021
20. Croatian Maritime Code, Official Gazette, no. 181/04, 76/07, 146/08, 61/11, 56/13, 26/15, 17/19
21. Letter to President of CMI (2021) CNII International Working Group on Vessel Nomenclature, Questionnaire #1 (March 2016), Comments of Dr. Bulent Sozer, Turkey, https://comitemaritime.org/work/ship-nomenclature/. Last accessed 25 Apr 2021
22. Reply by The Brazilian Maritime Law Association ("ABDM") to the CMI Questionnaire of March 2017 on the Study Relating to Unmanned Ships, Final Version, 29 May 2017. https://comitemaritime.org/work/mass/. Last accessed 5 May 2021
23. Jones Day (2017) White paper "Legal issues related to the development of automated, autonomous, and connected cars", November 2017. https://www.jonesday.com/en/insights/2017/11/legal-issues-related-to-the-development-of-automated-autonomous-and-connected-cars. Last accessed 1 July 2021
24. Coito J (2021) Maritime autonomous surface ships: new possibilities—and challenges. In: Ocean Law and Policy, 97 Int'l L. Stud. Ser. US Naval War Col. 259 (2021)
25. Allen C (2018) Determining the legal status of unmanned maritime vehicles: formalism v. functionalism. J Marit Law Commerce 49
26. Council Directive (1985) 85/374/EEC of 25 July 1985 on the approximation of the laws, regulations and administrative provisions of the Member States relating to liability for defective products. OJ L 210:29

Possibilities of Using Autonomous Green Vessels for Passenger Transport in Urban Environments

Mirko Čorić and Mladen Nikšić

Abstract In the cities around the world, road congestions affect the waste of time, increase of harmful gas emissions, increase of costs and increase of the likelihood of accidents. Conventional solutions for such problems understand usually the construction of new infrastructure, which is often expensive, time/space consuming, and insufficiently harmonised with the long-term goals of reducing the emission of greenhouse gases. Innovations, new smart and green technologies, as well as automation are the crucial factors in overcoming these modern challenges of mobility and logistics in urban environments. One of the modern innovative solutions to these problems lies in the exploitation of the sea and waterways of coastal cities, i.e. cities that lie on the shores of the seas, rivers and lakes. Autonomous green vessels in urban environments have the potential to become long-term and sustainable solutions that, apart from the revitalisation of the urban water transport, will enhance the economic, ecological, safety, and even tourist aspects of the cities. The paper gives a short insight into the technology of autonomous ships and their classification, and the specifics that result from their usage in urban environments. It gives an overview and analysis of the current condition of the sea/water transport in some of the famous world cities and an overview of the currently leading state-of-the-art projects and vessels suitable for urban environments. Finally, it analyses the possibility of using autonomous green vessels on the example of the city of Split (technical and technological perspective).

Keywords Urban water transport · Autonomous green vehicles · Linear shipping line systems · Environmental protection

M. Čorić
Faculty of Maritime Studies, University of Split, Ruđera Boškovića 37, 21000 Split, Croatia
e-mail: mirko.coric@pfst.hr

M. Nikšić (✉)
Faculty of Transport and Traffic Sciences, University of Zagreb, Vukelićeva 4, 10000 Zagreb, Croatia
e-mail: mladen.niksic@fpz.unizg.hr

M. Petrović et al. (eds.), *The Science and Development of Transport—ZIRP 2021*,
https://doi.org/10.1007/978-3-030-97528-9_7

1 Introduction

Many cities around the world have recognised this potential and directed their initiatives towards the mentioned solutions [1]. Many such popular solutions often relate to the establishment of regular shipping passenger lines with several stopping destinations within the urban environments [2]. The majority of the mentioned solutions in urban environments still do not include the aspect of environmentally friendly "green" and autonomous navigation, and the mentioned concepts are currently in their very infancy led by Norway as the pioneer of such solutions (more information further in the text).

The role and the importance of maritime and water transport in urban environments are reflected through the following several facts. About 40% of world population lives near the coast, and many major world cities are located near the sea, rivers, or lakes [3]. Eight out of ten largest world cities are coastal, and most of the non-coastal cities lie along the rivers [4]. It is estimated that by 2050 seventy per cent of the world population will live in the cities [5]. It should be noted that even now the regular passenger shipping lines transport annually around 2.1 billion passengers (for comparison, in air transport, the commercial flights carry annually around 2.3 billion passengers) [6]. For instance, Fig. 1 shows the annual number of carried passengers on regular shipping lines in several major world cities.

This type of transport in urban environments has been gaining increasing attention, and its further increase in future is expected (e.g. 20 million passengers a year in London by 2035, increase of carried passengers of about as much as 740% in San Francisco by the year 2035 compared to 2017) [8, 9]. The concept of autonomous green vessels in urban environments is a more flexible, more cost-effective and environmentally friendlier option in relation to the land infrastructure, bridges and conventional manned ships [3]. Autonomous green vessels are completely harmonised with the future trends of reducing greenhouse gases and the "smart cities" concept [10]. Such vessels can be harmonised with the future "smart" urban environment and intermodal transport system of the cities enhancing the quality of life of people in the cities. It should also be emphasised that the concept of autonomous vessels and the concept of "green" vessels are compatible to a great

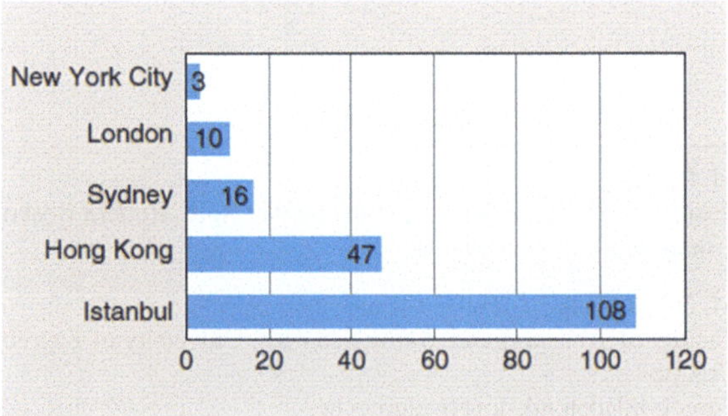

Fig. 1 Annual number of carried passengers on regular shipping passenger lines (in millions) [3, 7]

extent, and need not be considered separately. The majority of projects related to autonomous ships understands, namely, also the usage of renewable energy sources with inexistent or very low emission of greenhouse gases [11, 12].

2 Overview of Ship Passenger Lines in Major World Cities

The cities around the world are increasingly trying to solve the problems of traffic congestion and all the resulting negative consequences. The following overview of the level of development of the urban water traffic of some large world cities can best show its role and importance in urban traffic systems, as well as the potential of introducing autonomous green vessels in such environments. In the context of the researched topic, the overview of the cities and the related maritime/water traffic primarily presents general information about the existing urban water transport systems (the population of a single city, the number of carried passengers, the number of the existing ship lines and terminals, used vessels), as well as important information relevant for the future implementation of autonomous green vessels about the characteristics of the existing routes navigated at the moment by conventional vessels.

According to source [13] the navigation routes of urban environments can be classified into three basic types: (a)–(c). The routes of (a) type refer to routes where the ships navigate along the rivers or maritime/water channels within an urban area with several stopping destinations. Such lines are also called linear ferry systems [1, 2]. Routes of type (a) are extremely useful, maximising the efficiency and encouraging the development of coastal areas by providing stopping destinations along the coast. Such routes extend often parallel with land transport (e.g. along the rivers, maritime canals, etc.), and provide approximately equal trip times in peak periods of traffic congestion [14]. Typical examples of such routes can be found in Gothenburg in Sweden (Fig. 2).

Routes of type (b) refer to shorter routes with usually two (sometimes three) stopping destinations at river or sea crossings whose main purpose is to cross the water barrier in order to reach from one side of the shore to the other. This was in the past the most common form of ship passenger lines in urban areas developed primarily due to the lack or great distances of the land transport links and infrastructure (Fig. 3 shows an example of such routes in Amsterdam). Due to the short travel times, the design of the vessels on such routes is focused on manoeuvrability and capacity, whereas comfort is secondary.

Routes of type (c) refer to slightly longer routes that connect urban areas with the wider urban and suburban areas. Due to slightly longer trips, the frequency of ship departures on these lines is lower, and the design is focused on comfort (Fig. 4 shows an example of one such route in Stockholm about 25 km long that connects the peripheral area of Ekero with the centre of Stockholm).

Based on research [13] that was conducted in 23 world-famous cities with the existing shipping line transport systems, type (b) routes account for 43%, type (a) routes 37%, and type (c) routes 20%. Out of 23 cities, currently only five cities analyse

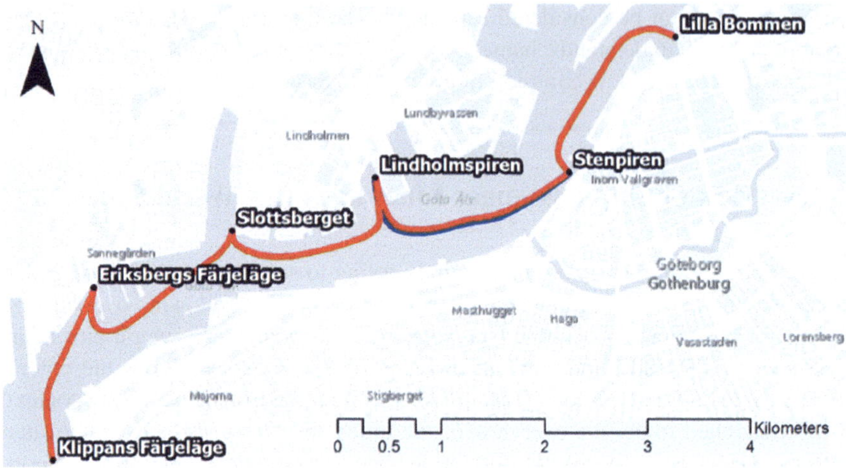

Fig. 2 Example of route type (a) in Gothenburg [13]

Fig. 3 Example of routes type (b) in Amsterdam [13]

the possibilities of introducing "green" vessels or autonomous vessels. Further is an overview of some of the best-known and largest urban maritime liner transport systems [13]: Amsterdam's liner shipping system (information in Table 1) is integrated into the city's multimodal transport system that, apart from vessels includes also the subway, trams, and buses, and a joint passenger ticket system. The vessels used are catamarans with a capacity (15 in total) of 150 passengers (Fig. 5). Since 2017, Amsterdam has been exploring the possibility of using autonomous vessels that would supplement and eventually replace the existing fleet of ships.

Furthermore, the Sydney Australian Shipping Line system (information in Table 2) is also integrated into the city's multimodal transport system (rail and bus transport) with a joint system of passenger tickets. Although the system is equipped with 28

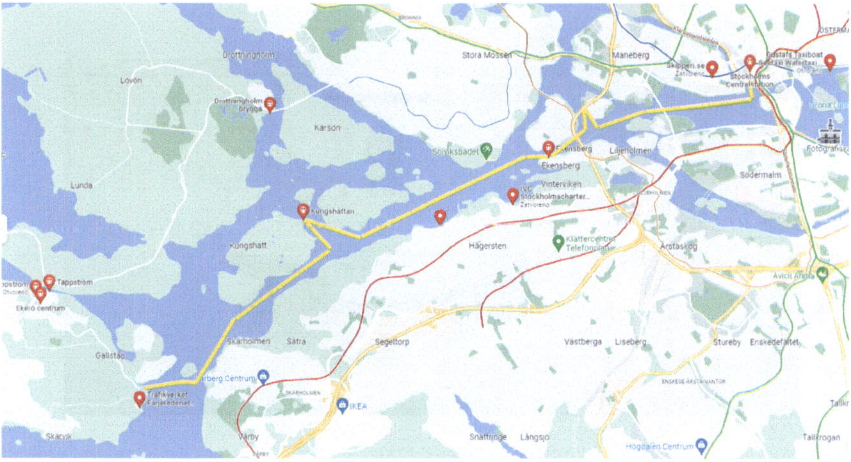

Fig. 4 Example of route type (c) in Stockholm (yellow line) [13]

Table 1 Ship line traffic system—Amsterdam [13]

Population	Urban area (km^2)	Types of routes	Number of routes	Number of piers	Number of passengers (year)
842,343	350	a, b	9	15	

Fig. 5 Catamarans sed in ship line transport system of Amsterdam [15]

Fig. 6 Autonomous electric vessel "miliAmpere" [23]

Table 2 Ship line traffic system—Sydney [13]

Population	Urban area (km^2)	Types of routes	Number of routes	Number of piers	Number of passengers (year)
4,921,000	12,367	b, c	8	39	14,700,000

obsolete slower monohull ships of a capacity of 300–400 passengers, the passenger satisfaction with shipping in Sydney is greater than the satisfaction with rail and bus transport.

One of the largest such transport systems in the world is in Venice (information in Table 3). The system is integrated into the multimodal transport system of the city, and consists of even 99 ships and 67 destinations. The ships used are monohull of a capacity of 210 passengers.

The water transport system of London (information in Table 4) is partly integrated into the rest of the city transport system, and the navigation is operated by 15 ships (catamarans) of a capacity of 150 passengers.

One of the cities with the most carried passengers on ship line transport system is Istanbul (information in Table 5). The system is not integrated into other modes of urban transport. Navigation is operated with 38 ships (monohull and catamarans) of the capacity of 600–1200 passengers.

Table 3 Ship line transport system—Venice [13]

Population	Urban area (km^2)	Types of routes	Number of routes	Number of piers	Number of passengers (year)
264,579	414	a	24	67	55,000,000

Table 4 Ship line transport system—London [13]

Population	Urban area (km^2)	Types of routes	Number of routes	Number of piers	Number of passengers (year)
8,673,713	1572	a, b	7	28	4,300,000

Table 5 Ship line transport system—Istanbul [13]

Population	Urban area (km^2)	Types of routes	Number of routes	Number of piers	Number of passengers (year)
14,025,646	1539	a, b, c	7	14	100,000,000

Ship line transport system of San Francisco (information in Table 6) expects a significant increase of demand in future. The system is currently not integrated into the multimodal transport system of the city. Navigation is operated by some thirty ships (monohulls and catamarans) of the capacity of between 150 and 788 passengers. Istanbul is researching the possibilities of changing the operation of the existing fleet from fossil fuels to renewable energy sources, as well as supplementing the existing fleet with new environmentally friendly "green" ships.

Ship line transport system of the city of Stavanger in Norway (information in Table 7) is integrated into the multimodal transport system of the city, and it consists of 24 ships (monohulls and fast catamarans) of the capacity of 180–296 passengers. A gradual transition to environmentally friendly "green" vessels is also planned.

Ship line transport system of the city of Stockholm (information in Table 8) is partly integrated into the multimodal transport system of the city, and consists of 11 older ships (monohulls) of the capacity of 225–300 passengers. The city is currently planning to increase the number of routes, as well as the number of operative ships.

Table 6 Ship line transport system—San Francisco [13]

Population	Urban area (km^2)	Types of routes	Number of routes	Number of piers	Number of passengers (year)
864,816	600	a, b, c	8	11	5,317,622

Table 7 Ship line transport system—Stavanger [13]

Population	Urban area (km^2)	Types of routes	Number of routes	Number of piers	Number of passengers (year)
130,426	71	b, c	9	46	4,200,000

Table 8 Ship line transport system—Stockholm [13]

Population	Urban area (km^2)	Types of routes	Number of routes	Number of piers	Number of passengers (year)
925,934	188	a, b, c	4 + 2 (test)	21	3,200,000

Table 9 Ship line transport system—Hamburg [13]

Population	Urban area (km^2)	Types of routes	Number of routes	Number of piers	Number of passengers (year)
1,774,242	755	a, b	7	22	8,000,000

The city of Hamburg owns a ship line system (information in Table 9) which is integrated into the multimodal transport system of the city. Twenty-four ships (monohulls) of a capacity of 200–400 passengers are operative on the routes. Hamburg is currently attempting to modernise the fleet, and is investigating the possibility of using completely "green" vessels driven by hydrogen fuel cells.

3 Autonomous Ships and Urban Transport

3.1 General Information and Challenges

Autonomous ships are becoming a reality, and according to a recent report and research of the Norwegian Shipowners Association, about half of the maritime companies will be using autonomous ships by the year 2050, while at the same time the Rolls-Royce Company plans to establish the navigation completely by autonomous ocean-going ships by 2035 [16]. Accordingly, the International Maritime Organisation has begun to develop the rules and regulatory frameworks that will refer to autonomous ships. According to the definition, autonomous navigation refers to the ability of a ship to independently control its own actions during the transport of goods or passengers from one place to another [17]. According to sources [17, 18] there are four basic types of automation:

1. Conventional ships with automated decision-support systems (e.g. collision avoidance system);
2. Partially autonomous ships in case of which the autonomous mode of operation of the ship is activated occasionally (e.g. during clear and stable weather);
3. Fully autonomous ships with crew that takes control and manages the ship during leaving and entering the ports;
4. Fully autonomous unmanned ships.

It is considered that in order to achieve maximum benefit (economic, environmental, and sociological) it is necessary to strive to develop the solution ad 4, i.e. fully autonomous unmanned ships [16]. It should be noted that according to the current research the long-term financial viability of using autonomous ships has already been confirmed (precise and detailed calculations of the financial profitability are not yet known [19]) although the initial costs are significantly higher than the use of standard ships [11]. Autonomous ship propulsion systems powered entirely by electricity without greenhouse gas emissions are the leading subject of research in most of today's projects for the development of autonomous ships [3]. The electric propulsion systems of most autonomous ships rely on fully self-contained battery power or fuel cells (hydrogen is the most commonly used as fuel) as source of electricity combined with batteries. The fuel cells and batteries do not pollute the environment, do not produce noise, and require much less maintenance than the conventional systems. Batteries and fuel cells are designed modularly, which renders them practical, easy to maintain and replace, and suitable for scalable design. Despite numerous benefits, the implementation and integration of autonomous ships into maritime and water transport will result in a number of general challenges and characteristics that maritime transport has not encountered so far [20]:

- *Scalable solutions*—requires a scalable design of autonomous ships that are adaptable to specific requirements.
- *Safety and acceptance by the passengers*—a matter of safety and acceptance of unmanned vessels by the passengers.
- *Navigation*—the way the navigation system works (navigation system should be predictable in compliance with the rules of navigation at sea, the so-called ColRegs rules of navigation).
- *Other vessels*—during navigation autonomous ships should react to other vessels, vessel facilities, swimmers, kayakers, buoys, etc. in a safe and predictable way. Currently, risk analysis methods and safety management methods for autonomous ships are currently being developed.
- *Land infrastructure*—infrastructure is needed for automatic docking, embarkation and disembarkation of passengers, and battery charging (battery technology is currently dominant).
- *Supervision and control*—advanced traffic monitoring is required, which includes remote monitoring and the possibility of managing autonomous vessels by means of specific remote control centres.
- *Cybersecurity*—this is the key to solving future problems of cyber-attacks and hacking, as well as software and communication vulnerabilities. This is one of the most important aspects that need to be considered prior to widespread use of autonomous ships.
- *Rescue systems*—strategy and method of rescue in the event of accident (collision, stranding, fire, etc.).
- *Testing*—long-term thorough testing of autonomous ships in real conditions is required before widespread use.
- *Regulatory and legal frameworks*—it is necessary to establish new rules that will regulate the navigation of autonomous ships (national and international).

3.2 Challenges in the Context of Urban Transport

In relation to the above-mentioned general challenges and characteristics of the use of autonomous ships, their use in urban environments brings some additional specifics and features:

- *Cybersecurity*—this aspect in urban fully networked and densely populated environments becomes even more important and an even greater challenge (e.g. exposure of an overseas autonomous ship to cyber-attacks in the middle of the ocean is lower than the exposure of an autonomous ship navigating in urban areas).
- *Characteristics of urban area routes*—it is necessary to take into account the fact that the routes within the urban areas are generally shorter and less complex than the standard maritime routes. As already mentioned, the majority of the routes within the urban environments belong to type (a) and (b) (see Chap. 2). This has several positive implications:
  - Adequacy of batteries as the only source of electricity on short routes.
  - Economic viability due to lower costs compared to the costs of the construction and maintenance of bridges or costs of manned ships (particularly on type (b) routes) [3].
  - Facilitated implementation of autonomous navigation and accompanying technologies such as artificial intelligence and algorithms for successful navigation due to the short distance of the route and often low traffic load of such sea/waterways (especially on type (b) routes), and generally fewer variables that need to be considered in the realisation of autonomous navigation on such routes [21].
- *Infrastructure*—increased number and price of port and terminal infrastructure of autonomous ships in urban environments (especially on type (a) routes).

3.3 Overview of Current Projects and Initiatives for the Development of Autonomous and Green Navigation in Urban Areas

The first project of developing an autonomous ship was the so-called MUNIN (Maritime Unmanned Navigation through Intelligence in Networks) started several years ago in Norway in cooperation with several European countries with the goal of enhancing the competitiveness and sustainability of the European maritime industry [11]. Autonomous green ships adapted to solving the problems of urban environments are a completely new concept with few such projects, the majority of which are still in the initial phase of realisation [3]. What follows is an overview of several such projects:

The Norwegian University of Science and Technology (NTNU) initiated the project "Autoferry" as a platform for research and development of autonomous navigation. The basic goal of the project is the development, design and construction of a fully autonomous "green" passenger vessel for safe and simple transport of passengers within the city of Trondheim. As part of the project, a prototype of an autonomous passenger vessel has been already developed called "miliAmpere" which is at the moment performing test sailings, and its full implementation and use are expected in 2021 in the navigation canal of the city of Trondheim. The capacity of vessel "miliAmpere" is 12 passengers, and it uses electric propulsion system and battery supply, and it is intended for crossing of the sea channel located in the area of the city of Trondheim. The concept is designed as an "on-demand ferry". The hull of the vessel is made of aluminium, and the power of the electric propulsion is 2 × 2 kW. The navigation systems Real-Time Kinematic (RTK), Global Navigation Satellite System (GNSS) and Inertial Measurement Unit (IMU) have been successfully implemented and tested. In case of failure of the mentioned systems the Simultaneous Localisation and Mapping (SLAM) was implemented and tested [22]. The system for dynamic positioning, automatic docking, remote control and collision avoidance system was also successfully implemented and tested on the prototype.

In 2016 two institutes, the Amsterdam Institute for Advanced Metropolitan Solutions and the Massachusetts Institute of Technology (MIT) launched the Roboat project [24, 25] which aimed to develop a fleet of smaller passenger vessels in congested urban areas. Currently, the tests with prototype vessels are being carried out in the network of sea canals of the city of Amsterdam. The latest prototype called "Roboat II" uses advanced algorithms for collision avoidance and navigation, space mapping algorithms, artificial intelligence, and it is equipped with numerous sensors, microcontrollers and cameras that achieve intelligent mobility on the waterways. The final version of the vessel has been designed as a floating autonomous platform (electric battery powered) with a capacity of 4 to 6 passengers, and it will also have the functionality of autonomous waste treatment in the water and water and air quality control via installed specialised sensors. Passenger transport will be performed by the "on-demand" concept so that the passenger requests transport at a certain location, and the nearest free vessel performs the transport with a flexible choice of route to the passenger's destination depending on the current traffic conditions on the waterways. Figure 7 shows the test prototype "Roboat II".

As part of the SVAN (Safer Vessel with Autonomous Navigation) project, Rolls-Royce Company, in cooperation with the Finnish Company Finferries built a fully autonomous commercial ferry called "Falco" (Fig. 8) for car and passenger transport, and in 2018 demonstrated fully autonomous navigation in Finland, between the cities of Parainen and Nauvo (in the Turku archipelago) [26, 27]. In addition to successful testing of the advanced technologies for collision avoidance and navigation, the success has also been achieved in the automated docking of this ferry with no additional human intervention. The ferry is 53.8 m long, 12.3 m wide, and a capacity of 54 cars. Although the ferry "Falco" does not belong to the category of fully green (zero-emission) solutions, it should be noted that the green solution of this ferry with its dimensions and capacity is already fully possible. This is proven by the current

Fig. 7 Autonomous electric vessel "Roboat II" [25]

Fig. 8 Autonomous ferry "Falco" [27]

two projects "Yara Birkeland" and "ReVolt" which cover the final construction phase of two commercial freight fully autonomous ships of similar dimensions, capacities, and range using electric propulsion and batteries [27–29].

The project "The Urban Water Shuttle" started in 2017 with the collaboration of several research centres and industrial partners involved in the business organisation the Norwegian Centres of Expertise Maritime CleanTech (NCE MCT) [30–32]. Initially, the project has the goal of developing green vessels (in future autonomous as well) that will help reduce urban congestion, greenhouse gas emissions, and costs of urban infrastructure, which will be applicable and adaptable (different modules) for the majority of cities with sea and waterways. The tests have been currently planned in the Norwegian cities, and commercial navigation is expected in 2022 in the city of

Stavanger (Norway). A prototype vessel with electric propulsion and battery drive, 25–30 m long, with capacity of 150–200 passengers and maximum speed of 20 knots is currently under construction. The prototype vessel will have a system of automated docking, and simultaneous charging of batteries during passenger reloading with an installed system of solar panels for additional battery supply (Fig. 9).

Based on the success of the "Autoferry" project and the "miliAmpere" vessel of the Norwegian University NTNU, the "Zeabuz" company will start in 2021 with commercial navigation of the autonomous green electric vessels in the city of Trondheim (Norway) which will operate on the basis of "on-demand" principle analogous to the "miliAmpere" vessel [33, 34]. The vessels will have a capacity of 12 passengers, the ride will take about one minute (compared to detour walking of about 15, 20 min), and the service will be free for passengers (Fig. 10). The batteries of the vessel will be charged at the docks.

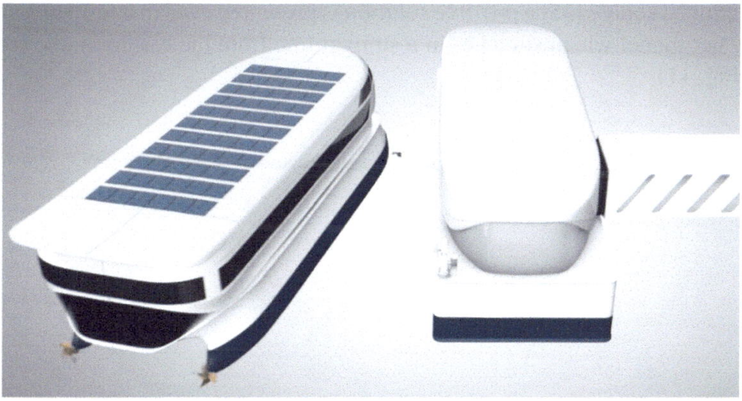

Fig. 9 Electric vessel "Urban Water Shuttle" [31]

Fig. 10 Autonomous electric vessel of the Norwegian Company "Zeabuz" [34]

4 Possibility of Using Autonomous Green Vessel in the City of Split

The traffic jams in road traffic are a significant problem of the city of Split, and with the growth of tourism and the extension of the tourist season the problem has deepened further. The influx of a large number of foreign tourists in the summer months results in the occurrence of large traffic jams in the area of the City Port of Split, and especially the roads on its eastern coast. Those passengers arriving on the eastern coast (city ferry port and bus station) need transport to the western coast of the Port City of Split, and also to other western parts of the city of Split. The transportation is done by road (usually taxi, rent-a-car, passenger cars) by detours in an arched path around the city centre (road distance of 4–6 km). Traffic connection of the eastern and the western coasts of the City Port of Split has been recognised as a solution that can reduce road congestion on the roads of the city of Split [35]. In the mentioned source [35] a possible solution is presented even in a form of a shorter underwater tunnel which stretches in a straight line from the eastern to the western coast (Fig. 11).

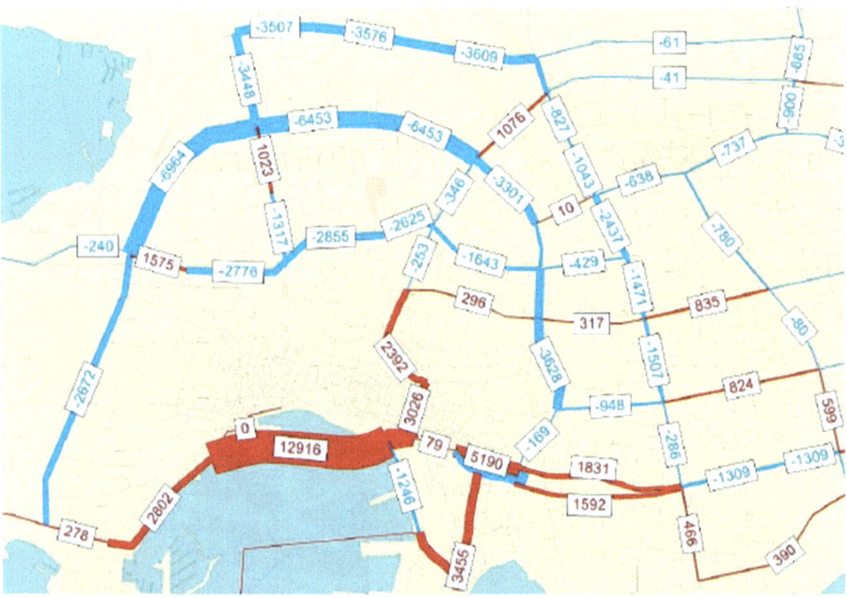

Fig. 11 Connecting the eastern and western coasts of the City Port Split by underwater tunnel and reduction of traffic volume [35]

The blue roads are the roads on which the volume of traffic would be reduced in case of connecting the eastern and western coasts. Negative numbers represent annual (traffic in 2015) reduction of the number of vehicles on individual roads that would follow the traffic connection of the eastern and the western coasts. The red roads and the respective positive numbers represent annual increase of the number of vehicles on the market roads that would follow in that case (at the level of 2015 traffic).

Further in the chapter, the possibility has been studied of alternative solution in the form of autonomous green "on-demand" vessels that would transport passengers between the eastern and the western coasts, and contribute to a certain extent to the reduction of congestion on the roads of the city of Split that are under the greatest load (especially in summer during the tourist season).

The possibility of using an autonomous green vessel in the city of Split has been studied in the context of connecting the eastern and the western coasts of the city port by using the existing solution created within the project of "Autoferry" and the associated vessel "miliAmpere", and the autonomous electric vessel of the Norwegian company "Zeabuz" (see the previous chapter). As described in the previous chapter, the designed concept is similar to the elevator concept, i.e. on-demand sailing. The vessel is intended for the crossing of the 95 m wide sea channel located in the centre of the city of Trondheim.

Technical specifications of the future commercial version of this vessel in their full size are described in paper [36] and listed in Table 10.

Proposal of the route in the city port for the navigation of the autonomous green vessel is shown in Fig. 12.

The route shown in Fig. 12 extends along the coast as illustrated due to the least impact on other ship traffic in the port. The route is about 950 m long and the vessel crosses it in 8 min if it sails at a speed of 4 knots. If the autonomous vessel had

Table 10 Technical characteristic of the future commercial version of the "miliAmpere" vessel [36]

Type of vessel	Autonomous passenger vessel
Maximum number of passengers	12
Number of seats	12
Number of places for bicycles	12
Number of places for the disabled	3
Length	9 m
Width	4 m
Vessel draught	0.515 m
Vessel displacement	7.610 t
Speed	4 knot
Main drive and energy supply	2 × PBES 400 v, 26 kWh battery modules

Fig. 12 Proposal of the route of the autonomous green vessel

25 kW of installed power (propulsion of 20 kW and other consumers 5 kW), after one crossing the vessel would consume 3.33 kWh of energy. Based on the example of the vessel of technical characteristics from Table 10, the autonomous vessel can make seven crossings and then need battery charging. For example, if determined that the charging time was 15 min, the power of the installed charging station would have to be in that case 104 kW, whereas for the charging time of 10 min the power of the charging station would have to be 156 kW. Charging after each crossing is also one of the possible solutions. For instance, if charging were done after each crossing and take 5 min (approximate time of passenger embarkation and disembarkation), the power of the charging station would have to be 40 kW. An example of a 150 kW car charging station ("Delta" company) occupies an area smaller than one square metre [37] and 40 kW charging stations of the company "Heliox" occupy a quarter of a square metre [38].

5 Conclusion

The population growth and the increase of economic activity in urban areas cause increasing traffic congestion, air pollution, noise, reduced road safety, and a general decline in the quality of life of the citizens. A large part of world population lives in urban areas located near the sea, rivers, and lakes, with the availability of numerous potential seas and water transport capacities, that are currently mostly underdeveloped. Consequently, the relocation of the traffic load to sea and waterways with the integration of such transport mode into multimodal transport systems of the cities poses itself as a logical solution to the mentioned problems of the cities.

This is supported also by the fact that sea/water transport, even with the use of the existing technology, represents a form of transport with the lowest greenhouse gas emissions. In addition, the mentioned solution would reduce the need for the demanding and expensive infrastructure interventions in road traffic. The use of the latest technologies and environmentally friendly solutions can be considered an imperative in the development of modern cities, and this paper studies the possibility of using autonomous green vessels for passenger transport in urban areas. The research showed that there are only few projects led by Norway, which include autonomous green vessels for passenger transport in urban areas, but also in general (currently there is a larger number of projects with autonomous green ships for freight transport). The reason can be found in the still insufficiently studied safety aspects and the acceptability of such transport by the passengers. The majority of the current projects is still in the test phase, and for the moment understand smaller vessels, narrower city cores and shorter simpler routes (type b). It is to be expected that the number of projects, the complexity of the routes and the capacity of vessels will grow in future. Apart from the reduction of traffic congestion and pollution, the autonomous green navigation has the potential of bringing savings to companies and passengers, and of enhancing the economic and tourist development of the cities. The implementation of the existing autonomous green vessel "miliAmpere" created as part of the "Autoferry" project has been analysed from a technical and technological perspective on the example of the city of Split. In the context of this perspective, the solution is justified, just as in the context of the need of the city of Split to connect the western and the eastern coasts of the City Port of Split. The implementation of this technologically advanced solution would reduce the use of taxi and rent-a-car vehicles, as well as passenger cars in the area of the City Port of Split and the surrounding roads, and the described solution would consequently contribute to the reduction of traffic jams and the reduction of air pollution.

References

1. Tanko M, Burke MI (2017) Transport innovations and their effect on cities: the emergence of urban linear ferries worldwide. Transp Res Procedia. 25:3957–3970
2. Thompson R, Burroughs R, Smythe T (2006) Exploring the connections between ferries and urban form: some considerations before jumping on bord. J Urban Technol 13(2):25–52
3. Reddy NP et al (2019) Zero-emission autonomous ferries for urban water transport: cheaper, cleaner alternative to bridges and manned vessels. IEEE Electrif Mag 7(4):32–45
4. The Explorer (2020) Official marketplace for green tech from Norway. Moving public transport onto waterways. Available at: https://www.theexplorer.no/solutions/urban-water-shuttle--moving-public-transport-onto-waterways/. Citirano 4 October 2020
5. Seto KC, Shepherd JM (2009) Global urban land-use trends and climate impacts. Current Opin Environ Sustain 1(1):89–95
6. Zanne M, Twrdy E, Zlak B, Stojaković M (2016) Development possibilities for ferry transport in Adriatic-Ionian region. Ekonomiczne Problemy Usług 124:39–52
7. Charles Carter—Smart Cities World. Autonomous passenger ferries: Congestion-buster or hype on the high seas? Available at: https://www.smartcitiesworld.net/special-reports/special-

reports/autonomous-passenger-ferries-congestion-buster-or-hype-on-the-high-seas. Citirano 5 Listopad 2020
8. Gougherty M (2020) Pacific Maritime Magazine. WETA Plans major expansion of ferry service on the SF Bay. Available at: https://www.pacmar.com/story/2017/10/01/features/weta-plans-major-expansion-of-ferry-service-on-the-sf-bay/552.html. Citirano 05 Listopad 2020
9. Khan S (2020) London's passenger pier strategy—a safe, sustainable and integrated pier network for London. Available at: http://content.tfl.gov.uk/pier-passenger-pier-strategy-action-plan.pdf. Citirano 5 Listopad 2020
10. Kaluarachchi Y (2019) Improving urban city resilience by incorporating smart green infrastructure components. In: International Council for Research and Innovation in Building and Construction (CIB) 2019, Hong Kong Polytechnic University
11. Munim ZH (2019) Autonomous ships: a review, innovative applications and future maritime business models. Supply Chain Forum Int J 20(4):266–279
12. European Union. Smart City Ferries—solutions and opportunities. Available at: https://www.aboamare.fi/_media_25440_/R&D%20kuvat/%c3%84lyVESI/Tulokset%20201805/Smart-City-Ferries-Solutions-and-Opportunities.pdf
13. Cheemakurthy H, Tanko M, Garme K (2017) Urban waterborne public transport systems: an overview of existing operations in world cities. Technical report. KTH Royal Institute of Technology
14. Tanko M, Burke MI, Yen B (2019) Water transit and excess travel: discrete choice modelling of bus and ferry trips in Brisbane, Australia. Transp Plan Technol 42(3):244–256
15. Travelguide Amsterdam (2020) Ferries in Amsterdam: info, tickets, lines & tips for passengers. Available at: https://www.travelguide.amsterdam/en/publictransport-gvb/gvb-ferries/. Citirano 13 Listopad 2020
16. Amro A, Gkioulos V, Katsikas S (2019) Connect and protect: requirements for maritime autonomous surface ship in urban passenger transportation. Computer Security, pp 69–85
17. Rødseth ØJ (2017) From concept to reality: unmanned merchant ship research in Norway. In: Proceedings of Underwater Technology (UT) 2017, IEEE, Busan, Korea, pp 1–10
18. International Maritime Organization—IMO (2020) IMO takes first steps to address autonomous ships. Available at: https://www.imo.org/en/MediaCentre/PressBriefings/Pages/08-MSC-99-MASS-scoping.aspx. Citirano 13 Listopad 2020
19. Blanke M, Henriques M, Bang J (2017) A pre-analysis on autonomous ships. Technical University of Denmark
20. Tannum MS, Ulvensøen JH (2019) Urban mobility at sea and on waterways in Norway. J Phys Conf Ser 1357(1):012018
21. Reactor (2020) Urban waterways: the next generation of autonomous transportation. Reactor. Available at: https://www.reaktor.com/work/autonomousferry/. Citirano 15 Listopad 2020
22. Ødven MS (2019) Lidar-based SLAM for autonomous ferry. Master's thesis. NTNU: Norwegian University of Science and Technology
23. NTNU (2020) Norwegian University of Science and Technology. Autonomous all-electric passenger ferries for urban water transport (Autoferry). NTNU: Norwegian University of Science and Technology. Available at: https://www.ntnu.edu/autoferry. Citirano 15 Listopad 2020
24. Wang W, Gheneti B, Mateos LA, Duarte F, Ratti C, Rus D (2019) Roboat: an autonomous surface vehicle for urban waterways. In: 2019 IEEE/RSJ International Conference on Intelligent Robots and Systems (IROS), pp 6340–6347
25. Massachusetts Institute of Technology (2020) Autonomous boats could be your next ride. Available at: https://news.mit.edu/2020/autonomous-boats-could-be-your-next-ride-1026. Quoted 16 Oct 2020
26. Felski A, Zwolak K (2020) The ocean-going autonomous ship—Challenges and threats. J Mar Sci Eng 8(1):41
27. Safety4Sea (2020) Rolls-Royce, Finferries present world's first fully autonomous ferry. Available at: https://safety4sea.com/rolls-royce-finferries-present-worlds-first-fully-autonomous-ferry/. Quoted 16 Oct 2020

28. YARA (2020) YARA selects Norwegian shipbuilder VARD for zero-emission vessel Yara Birkeland. Available at: https://www.yara.com/corporate-releases/yara-selects-norwegian-shipbuilder-vard-for-zero-emission-vessel-yara-birkeland/. Quoted 17 Oct 2020
29. Autonomous Ships HQ (2020) REVOLT. Available at: https://www.autonomousshipshq.com/revolt/. Quoted 17 Oct 2020
30. Norwegian Centres of Expertise Maritime CleanTech (NCE MCT) (2020) World-leading cluster for clean maritime solutions. Available at: https://maritimecleantech.no/about-us/. Quoted 17 Oct 2020
31. The Explorer (2020) Official marketplace for green tech from Norway. Moving public transport onto waterways. Available at: https://www.theexplorer.no/solutions/urban-water-shuttle--moving-public-transport-onto-waterways/. Quoted 25 Oct 2020
32. Norwegian Centres of Expertise Maritime CleanTech (NCE MCT). Urban Water Shuttle—The waterway to a sustainable future. Available at: https://maritimecleantech.no/wp-content/uploads/2014/12/UWS-folder.pdf
33. Zeabuz (2020) Zero emission autonomous urban and coastal mobility. Available at: https://zeabuz.com/. Quoted 26 Oct 26
34. Diskin E (2020) Matadornetwork. Self-driving, electric ferries are coming to Norway next year. Available at: https://matadornetwork.com/read/self-driving-electric-ferries-norway/. Quoted 26 Oct 2020
35. Split G (2011) Prostorno—Prometna Studija Šireg Područja Grada Splita. Available at: https://es.split.hr/Documents/2018/08/23/Knjiga%201.pdf
36. Havdal G, Heggelunnd CT, Larssen CH (2017) Design of a small autonomous passenger Ferry. Master's thesis. NTNU: Norwegian University of Science and Technology
37. Delta (2020) 150 kW ultra-fast charger. Available at: https://emobility.delta-emea.com/downloads/Brochure%20-%20UFC%20Ultra%20Fast%20Charger%20V7%20EN%202019-10-07.pdf.pdf. Quoted 18 Nov 2020
38. Heliox (2020) Mobile charger fast DC 40 mobile. Available at: https://www.heliox.nl/images/products/2020/FAST_DC_40_kW_mobile.pdf. Quoted 18 Nov 2020

State and Projection of the Port Traffic in the Port of Split: Post-COVID Era

Zvonimir Lušić, Luka Vukić, Danijel Pušić, and Ivica Skoko

Abstract This paper analyses the total port traffic in the Port of Split comprising both the passenger port basin, nominated as the City port, and cargo terminals situated in the northern part of the city and specified as the North port. The overview of the annual trend of traffic handled by individual port area is elaborated in detail. Considering the main objective of this research, the central part of the paper provides the breakdown of the overall ship throughput concerning the criterion of ship type and size together with the monthly distribution of arrivals for the two most important categories in the port, passenger and cargo ships. The traffic of tourist and other smaller vessels is also taken into account. The passenger and cargo ship traffic data from 2019 represents the basis of the analysis, with an indication that the overall traffic in 2020 is described separately mainly due to the impact of the COVID-19 pandemic on the decline in total port traffic. In addition to the presented statistical analysis, the final part of the paper provides an assessment of the ship traffic recovery in the Port of Split based on the application of linear trend projection. The obtained results confirm that a positive trend of ship traffic growth in the Port of Split may still be expected in years to come, regardless of the temporary decline in traffic caused by the COVID-19 pandemic. It can be expected that a positive growth trend will correlate primary with the passengers and cars traffic, as well as tourist activities within the wider port area.

Keywords Port of Split · Maritime traffic · Passenger ships · Cargo ships · COVID-19 impact

1 Introduction

The Port of Split is located in the middle of the east coast of the Adriatic Sea [1] and is the largest port in Dalmatia, as well as the largest passenger port in Croatia. Its importance is reflected by providing mobility for the local population to various

Z. Lušić (✉) · L. Vukić · D. Pušić · I. Skoko
Faculty of Maritime Studies, University of Split, Ruđera Boškovića 37, 21000 Split, Croatia
e-mail: zlusic@pfst.hr

M. Petrović et al. (eds.), *The Science and Development of Transport—ZIRP 2021*,
https://doi.org/10.1007/978-3-030-97528-9_8

destinations along the coast, connecting the Central Dalmatian islands with the mainland, and serving a large number of tourists, whether the port serves as a transit or final destination. The port area is divided into seven berthing areas, namely the city port basin (passenger and vehicle port), basin Vranjic-Solin (cargo port), Kaštela basin A, Kaštela basin B, Kaštela basin C, Kaštela basin D, Resnik and Komiža basin for fishing needs [2]. The Port of Split as a system is divided into two separate components, a passenger terminal located in the city centre (City Port) and cargo terminals located in the northern part of the city (North Port) [3].

Despite having significant and positive trends of cargo throughput volumes in the recent years, the Port of Split is primarily perceived as the port for national and international passenger traffic [4]. The main and projected markets and gravitational area of the Port of Split are shown in Fig. 1.

Fig. 1 Main markets and gravitational area of the Port of Split [5]

The potentials of the long-term development of the Port of Split closely relate to the obligatory investments and modernization of maritime and rail infrastructure as a prerequisite for future growth in both passenger and especially freight throughput [4]. A more detailed overview of the characteristics and specifications of the Port of Split, especially in relation to cargo terminals, can be found in Vukić et al. [6]. Cargo traffic in the Port of Split is secondary important in relation to passenger traffic and is primarily used to satisfy the needs and demand of the local market. Traffic in the Port of Split has grown steadily over the last 20 years, a trend that peaked in 2019. The expected upward trend would likely have continued if the global crisis caused by the COVID-19 pandemic had not occurred in the first quarter of 2020. Figure 2 shows the annual trend in passenger and vehicle traffic, both domestic and international, along with cargo traffic activity at the Port of Split.

While the trend in the segment of cargo throughput in the Port of Split showed variations in the observed period (30 years), the data from Fig. 2 provides a linear increase in passenger and vehicle throughput during the years, which peaked in 2019. The Port of Split handled 5.6 million passengers and more than 800,000 vehicles in 2019 [7]. The growth of freight and passenger traffic goes in conjunction with the growth in the number of ships. The structure of shipping traffic, based on their

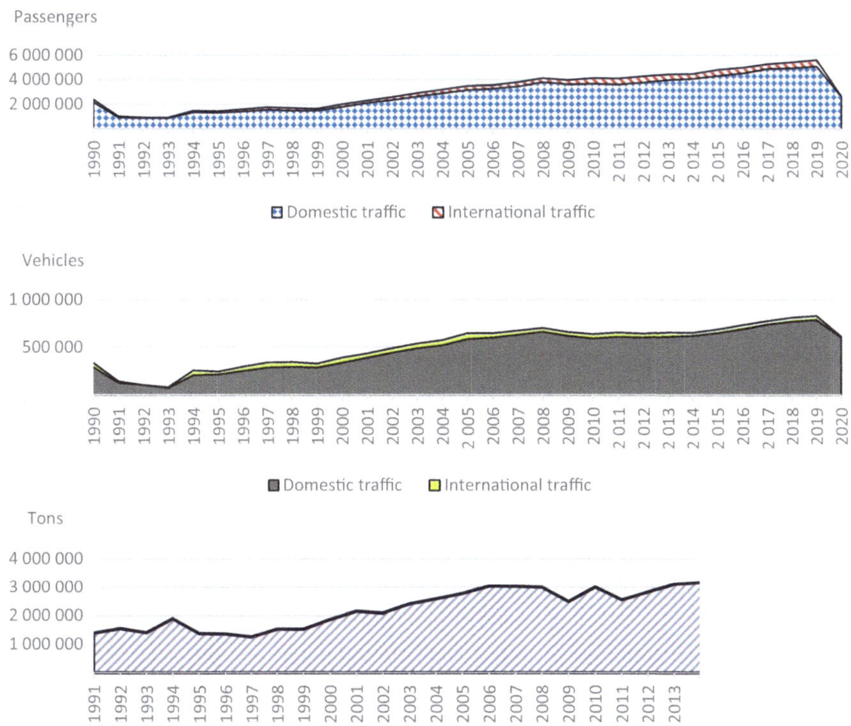

Fig. 2 Passenger and vehicle traffic (above graph) and total cargo traffic (below graph) in the Port of Split [7, 8]

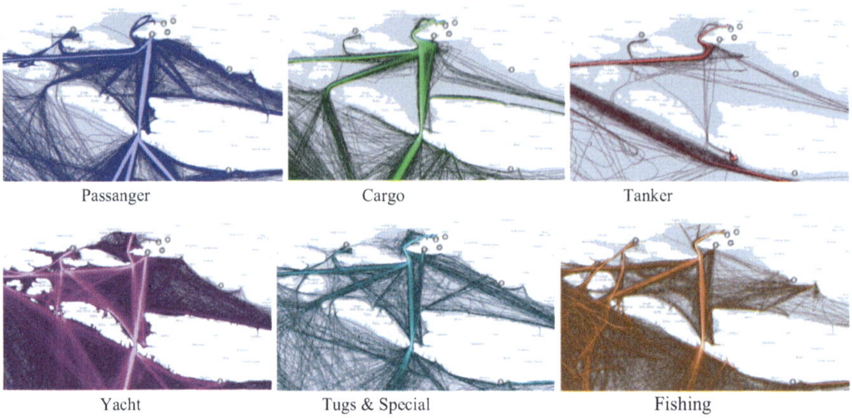

Fig. 3 Marine traffic on approach to the Port of Split (AIS data from 2017) [9]

arrivals and departures, will be analysed in more detail in later chapters. Figure 3 shows the spatial distribution and density of maritime traffic by type of vessel, when approaching the Port of Split.

The variety of ship types and proportionally types of cargo indicate the complexity of passenger and cargo operations in the Port of Split. This is primarily visible in the diverse passages and respective density of ships utilizing available main access fairways and routes to the port. Main access fairways to the City port include the Brač channel, Šolta channel, Drvenik channel and most frequently used passage Splitska vrata, while the significance of other coastal fairways is negligible [10]. Besides the predominant and consistent ferry passenger and other similar transport modes (catamaran, yacht and other tourist vessels), the cargo transport activities are based on the weekly feeder container service, almost regular tanker routes in the distribution of petroleum along the eastern Adriatic and other cargo ships with freight for and from the gravitational port area.

The objective of this paper is to provide the overview of the port traffic in the Port of Split, comprising both the cargo and passenger throughput and overall traffic based in the criterion of total ships handled. It should indicate the importance of the port activity to the broader community and highlight the importance in both economic and social component. The secondary activity includes the overview of smaller vessels traffic and its contribution to the port traffic activity. Additionally, the effects of the COVID-19 impact of the port traffic indicators in 2020 in correlation with the tourist activity on the local and regional level are shown. The future trends in transport activity on passenger and cargo terminals of the Split port are performed by the use of linear trend projection, which should determine the perspectives and recovery dynamics.

2 Ship Arrivals and Departures Indicators

In 2019 the Port of Split handled the total of 23,844 ships (arrivals and departures), of which 84.7% referred to the City port, and 15.3% to other port basins [11]. These indicated levels are slightly higher than those recorded in 2018 (20,382), while the values in 2020 showed a significant decline in total ships handled in the port (Table 1). Domestic ships predominate in total port traffic, participating in the share of 96% in 2019. Based on the type of domestic maritime transport services, the local ferry and catamaran cabotage activity provided the largest proportion of marine traffic in the City port in 2019 (70.8%), followed by tourist and excursion boats (26%). From the remaining traffic activity in the City port, 238 arrivals of ships in international liner traffic, 282 arrivals of cruise ships, 89 arrivals of other ships (tugs and work boats) and 52 arrivals of yachts [11] should also be mentioned. Details of ship and cargo traffic are provided in Table 1.

The North port, comprising the cargo terminals on the northern suburb of the city, is primarily utilized for handling cargo ships. During 2019, a total of 3657 ship arrivals were recorded in the Kaštela basins B, C and D, and in the Vranjic-Solin basin, where the share of 54.6% referred to the arrivals of cargo ships [11]. It should be emphasized that the arrivals of cargo ships from domestic ports predominate in the total port traffic in the North port, in the share of 68% [12]. The total cargo throughput in the Port of Split (considering all the cargo activities and port basins) in 2019 amounted to 2.6 million tons, the passenger traffic realized was at the level of 6.6 million passengers, and vehicles traffic close to 800,000 units. However, cargo traffic activity levels in 2020 have remained stable, while passenger traffic decreased to 2.5 million, and vehicle traffic to around 600,000 of units [13].

3 The Structure of Marine Traffic in the Port of Split

The total port traffic in the Port of Split has been steadily increasing in recent years, with the exception of 2020 and the recorded decline caused by the COVID-19 pandemic. Table 2 shows the trend ratio of marine traffic in the City port and other basins in a seven-year period.

The structure of port traffic activities in the City port, based on the criterion of the ship type, is shown in Table 3. The majority of traffic is realized by the local ferry cabotage and catamaran line traffic (70.1%), followed by national tourist vessels (15.9%) and excursion ships (10.0%).

The analysis of the structure of traffic in the City port, considering the ship size, shows predominant category of vessels, which relates to the ships up to 500 gross tonnage (GT) (47.3% on a sample of 18,666 entry records). The largest ship that arrived at the City port basin, during 2019, had 99,836 GT, and the length of 299.8 m. Vessels up to 500 GT (58.3%, on a sample of 1490 entry records) also prevail in the North port (Fig. 4) [14].

Table 1 Traffic in the port of split [12]

Marine traffic	2020		2019		2018	
	Split	Kaštela	Split	Kaštela	Split	Kaštela
Number of arrivals of domestic ships	12,712	21	22,868	177	20,658	149
Domestic ships from a foreign port	106	–	232	/	241	/
Domestic ships from a domestic port	12,606	21	22,636	177	20,417	149
Number of arrivals of foreign ships	762	43	1,019	19	1,063	21
Foreign ships from a foreign port	488	–	659	/	762	/
Foreign ships from a domestic port	274	43	360	19	301	21
Number of departures of domestic ships	12,713	20	22,871	183	20,648	149
– For a foreign port	106	–	233	/	241	/
– For a domestic port	12,607	20	22,638	183	20,407	149
Number of departures of foreign ships	764	43	1,021	19	1,062	21
– For a foreign port	490	–	659	/	762	/
– For a domestic port	274	43	362	19	300	21
Total boats (individual)—in foreign trade	496	–	700	/	790	/
Loaded cargo in tons	1,531,096	121	1,013,375	/	1,170,075	51
– Of which dangerous cargo	13,019	–	42,321	/	20,216	51
Discharged cargo in tons	1,098,094	–	983,361	/	934,931	/
– Of which dangerous cargo	738,810	–	655,137	/	611,894	/
Passengers embarked	1,289,290	28	3,362,387	148	3,586,152	2
Passengers disembarked	1,252,798	11	3,275,374	1489	3,499,350	112
Cars loaded	302,819	2	409,919	/	396,014	4
Cars unloaded	297,946	–	404,280	/	391,064	149

The largest cargo ship to enter the North Port in 2019 was 25,373 GT and 175.97 m in length. If the total annual traffic is converted to daily traffic, it results in 50 vessels arriving at the city port per day and about 6 vessels/day arriving at the North Port. It should be emphasized that the port traffic values analysed do not take into account the traffic of smaller vessels that contribute to the importance of the split port system. The actual daily level of vessel throughput in the Port of Split changes significantly throughout the year, with significantly higher values in the peak summer periods.

Table 2 Total arrival traffic in the Port of Split [7]

	City port	Basins	Total
2013	15,107	1115	16,222
2014	15,604	1100	16,704
2015	16,856	1086	17,942
2016	17,721	1474	19,195
2017	16,439	2107	18,546
2018	18,237	2145	20,382
2019	20,187	3657	23,844

Table 3 City port of split—the structure of marine traffic by ship types [7]

	2017	2018	2019
Ferry line domestic traffic	9686	13,372	14,284
Catamaran line domestic traffic	2424		
International line traffic	258	267	238
Visit of tourist vessels of the Republic of Croatia	2610	2984	3215
Visit of yachts	110	18	52
Cruise ships	234	260	282
Excursion ships	892	1231	2027
Seaplanes	/	/	/
Various ships (tugs, work, …)	225	105	89
Total	16,439	18,237	20,187

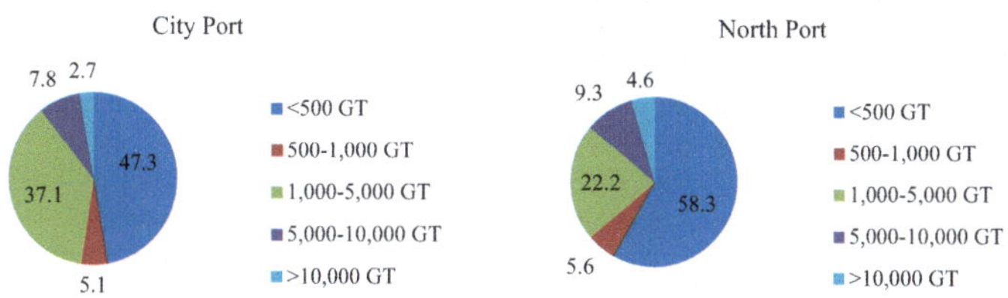

Fig. 4 The traffic structure by ship size for the City and North Port of Split [13, 14]

Figure 5. shows the monthly vessel traffic in the Port of Split, while Fig. 6. shows the daily comparison of traffic in the Port of Split in July and December 2019, based on the obtained AIS data for tracking the (larger) vessels. The data shows that the expected number of arrivals in the Port of Split in the low season is around 30–40 vessels, while port activities in the peak tourist season may be associated with the arrival of 90–100 vessels [13, 14].

By analysing the changes of the shipping timetable in peak and off season, i.e. the frequency of departures on ferry lines, the extent of the difference in total monthly

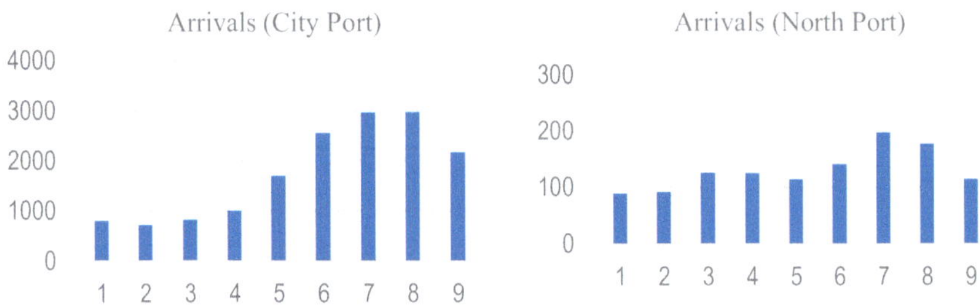

Fig. 5 Monthly traffic of the Port of Split in 2019 [13, 14]

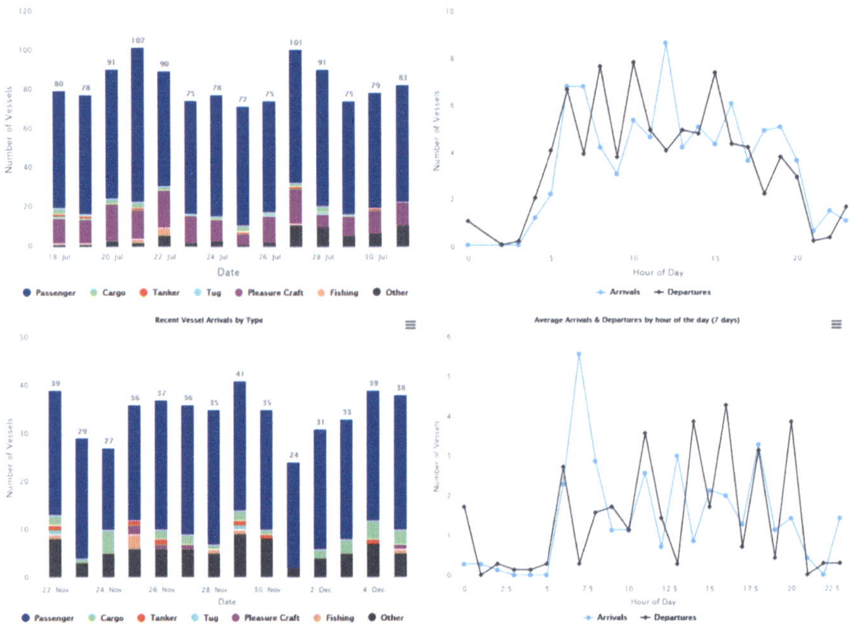

Fig. 6 Daily number of arrivals in the Port of Split—July/December 2019 [15]

traffic, i.e. increased number of arrivals and departures of ships during the summer season can also be highlighted (Table 4).

The local liner traffic frequency in the lower season is almost twice lower in comparison with the dynamics of the higher summer period, which can be associated with seasonality and strong dependence on tourist activity. Seasonal character of marine traffic in the Port of Split is also visible in the handling of cruise vessels (Table 5).

The majority of this type of traffic was realized in the period April–October. It should be noted that during 2020, there were almost no cruise ships due to the COVID-19 pandemic. Furthermore, the impact of COVID-19 on the overall port

Table 4 Local liner traffic in the City Port of Split [16–18]

Line	Season (daily departure)	Out of season	Ship
Split-Supetar	14/12	8–9	Ro ro pass. (Jadrol.)
Makarska-Sumartin (Brač)	5–4	3	Ro ro pass. (Jadrol.)
Split-Stari Grad (Hvar)	7	3	Ro ro pass. (Jadrol.)
Drvenik-Sućuraj (Hvar)	11/10/9	6/7/8	Ro ro pass. (Jadrol.)
Split-Rogač (Šolta)	6/5	4	Ro ro pass. (Jadrol.)
Split-Vis	3–2	2	Ro ro pass. (Jadrol.)
Split-Vela Luka-Ubli	4	2–1	Ro ro pass. (Jadrol.)
Split-Trogir-Drvenik V/M	4–3	3	Ro ro pass. (Jadrol.)
Split-Bol-Jelsa	1	1	HSC (Jadrol.)
Split-Milna-Hvar	1	/	HSC (Jadrol.)
Split-Hvar-Vela Luka-Ubli	1	1	HSC (Jadrol.)
Split-Hvar-Prigradica-Korčula	1	1	HSC (Jadrol.)
Split-Hvar-Korčula	1	1	HSC (Jadrol.)
Split-Milna-Hvar-Vis	1	1	HSC (Krilo)
Split-Brač-Hvar-Kor-Mljet-Dubr	1	0/3–4 weakly	HSC
Komiža-Porat (Biševo)	1	4 per weak	Passenger ship
Bura Line	9–4	/	Passenger ship

traffic was visible in the significant reduction of international liner traffic values, where a large number of passenger lines were cancelled, while the remaining lines were maintained to a lesser extent and with interruptions (Table 6).

The six international passenger lines handled only 38,244 passengers and 13,763 vehicles in 2020, which represents a decrease of 91.3% compared to the overall turnover of passengers realized and 78.8% decrease in total amount of vehicles compared to 2019 [19].

The activity of overall turnover of ships and cargo in the North port, more specifically in the port basins of the Kaštela Bay and the Vranjic-Solin basin, has also been characterized by the element of seasonality, but in overall it is much less pronounced. Table 7 shows the traffic of ships in port basins of the Port of Split.

The generated values in Table 7. show significant increase of almost 2.5 times during the observed period (2016–2019). Number of arrivals segmented according to port basins, for 2019, is shown in Table 8.

The variety of port activity in relevant port basins in evident, highlighting the industrial and tourist component in port portfolio and activity. The overview of the number of ship arrivals in diverse port basins can be further analysed with the intensity of the COVID-19 impact, where the port basins with strong industrial component sustained slight decrease, while the remaining port basins and terminals having the developed tourist activity suffered heavily, with significant decrease in port traffic levels.

Table 5 Tourists and cruise ships turnover [7]

	2014		2015		2016		2017		2018		2019	
	No. arr	No. tourists	No. arr	No. tourists	No. arr	No. tourists	No. arr	No. tourists	No. arr	No. tourists	No. arr	No. tourists
January	3	147	3	118	4	136	4	184	4	2,897	3	2,642
February	5	237	6	262	5	166	4	181	5	246	6	858
March	6	255	6	622	11	4,767	7	2,565	10	5,946	8	3,651
April	21	19,501	14	12,478	8	17,318	12	15,068	8	12,012	23	31 326
May	35	22,955	43	51,442 1 442	35	32,125	28	24,443	36	40,376	34	48 040
June	32	26,803	32	34,259	37	37,805	26	28,131	40	44,018	34	48 272
July	26	28,698	29	35,421	34	44,794	26	31,643	29	41,750	37	51 680
August	32	29,122	30	34,208	37	44,161	32	28,552	40	47,962	38	54 112
September	39	36,503	43	45,233	46	42,621	33	34,288	34	42,946	39	49 492
October	23	19,373	28	30,736	34	32,932	39	37,148	27	38,361	40	54 461
November	4	179	21	26,405	16	20,000	13	19,204	17	20,416	14	15 211
December	7	289	6	261	9	1,434	10	10,837	10	10, 218	6	210
Total	233	184,062	261	271,445	86	278,259	234	232,244	260	307,148	282	359,955

Table 6 Number of arrivals per month in international liner traffic [11]

	Month											
Total	I	II	III	IV	V	VI	VII	VII	IX	X	XI	XII
238	10	8	9	20	26	28	31	44	27	17	9	9

Table 7 Arrival of cargo and other vessels by basins in the period 2016 2019 [11]

	2016	2017	2018	2019
Vranjic-Solin basin (cargo) (yachts, tourist and others)	340 196	600	176 475	822
Kaštela basin B (cargo)	678	747	803	664
Kaštela basin C (cargo) Visit: arr/dep. Fishing boats Brižine	214	717	691	509
Kaštela basin D (cargo)	24	23	/	1.662
Total	1474	2107	2145	3657

Table 8 Number of ship arrivals in the port basins of the Port of Split in 2019 [11]

Port Basins	Activity
Vranjic-Solin basin	Berths no. 1 to 5—139 arrivals of cargo ships; Fishing boats (berth no. 1)—99 arrivals; Berth no. 6 (silos)—37 arrivals; Berth no. 1, 7/8 (tourist)—518 arrivals; Various ships—29 arrivals
Kaštela basin B	Coast of Sv. Juraj I / II—262 arrivals; Dangerous cargo (berth 5)—307 arrivals; Others—95 arrivals
Kaštela basin C	Coast of sv. Kajo (berth 1)—7 arrivals; INA tanker terminal—167 arrivals; Brižine (Fishing boats)—312 arrivals / departures; Various ships—23 arrivals
Kaštela basin D	Excursions boats—1.213 arrivals; Liner ships—449 arrivals

4 Traffic of Smaller Vessels

The traffic of smaller ships is very difficult to determine, as there is no systematic way to monitor them. Moreover, smaller vessels do not have fixed routes, and the intensity of their traffic is much higher in the summer months compared to the rest of the year. An approximate estimate of the traffic of smaller vessels can be made in terms of their number, i.e. the number of tourist vessel arrivals recorded. In 2019, the number of fishing vessels in the Republic of Croatia was 336 (with a total of 27,897 GT), and the number of boats was 7278 [20]. The number of fishing vessels whose home port was in Split was 107, and the number of fishing boats was 108 (in addition Kaštela 31) in 2019. The total number of boats with the port of registry in Split (together with Kaštela) was 8479 in 2018, and yachts 1,013. A detailed overview of the status of vessel registration for Split and Kaštela can be found in Table 9 [12].

Table 9 The state of ships with port of registry in Split [12]

The state of the Croatian fleet	2020	2019	2018	2017	2016	2015
Number of registered ships in total	540	583	549	534	551	539
– *Merchant ships*	469	465	432	420	403	395
Cargo	53	53	53	54	136	135
Passenger	350	347	314	306	217	212
Technical vessels	66	65	65	60	50	48
– *Fishing ships*	108	108	107	104	143	139
– *Public ships*	10	10	10	10	5	5
Number of ships under construction	27	15	39	42	46	41
Number of newly registered ships	4	34	48	30	15	23
Number of deleted ships	47	/	31	33	34	34
Number of registered boats in total	6176	7747	7594	7300	7309	7262
– *Boats for commercial purposes*	906	1091	949	661	638	614
For cargo transport	58	72	67	61	38	38
For passengers' transport	128	164	142	158	473	438
Fishing boats	57	115	108	114	127	138
For rent	653	740	632	328		
– *Boat for public purposes*	28	31	28	23	24	27
– *Boat for personal use*	4239	6755	6617	6616	6647	6621
Number of newly registered boats	223	280	419	474	412	401
Number of deleted boats	1764	127	125	152	342	393
Number of registered yachts in total	1070	1113	1013	879	814	756
– *Yachts for commercial purposes*	850	890	800	679	626	571
– *Yachts for personal use*	220	223	213	200	188	185
Number of newly registered yachts	27	145	186	144	114	80
Number of deleted yachts	70	45	57	49	56	61

On the other hand, the total number of nautical tourism ports in the Republic of Croatia in 2019 was 167, while in Split-Dalmatia County that number was 31. Considering the trend analysis in national level, it represents a 17.6% increase compared to 2018, while on the county level an increase of 6.9%. Capacities of nautical ports in the Republic of Croatia and in Split-Dalmatia County by port categories in 2019 are shown in Table 10.

The Split-Dalmatia County has had for the nautical tourism activity purposes at disposal a total area of 574,862 m^2, with a total of 2370 berths, and 753 places to accommodate vessels on land in 2019. The total land area was 95,032 m^2. The total water area in the Republic of Croatia, at the same time, was 4,349,400,270 m^2, and the total number of berths was 18,179. The number of vessels with a permanent berth in nautical tourism ports of Split-Dalmatia County in 2019 amounted to a total of

Table 10 Nautical tourism ports (2019) [21]

Territory	Total	Anchorage	Berthing	Marinas					Boat storage
				Dry marina	I. cat	II. cat	III. cat	Categorized and marked with anchors	
Croatia	167	75	9	17	5	13	17	26	5
Split-Dalmatia County	31	15	1	4	/	3	3	5	/

Table 11 Number of vessels with a permanent berth in nautical tourism ports (2019) [21]

	Total	Motor yachts	Sailing boats	Other
Total Croatia	14,249	7249	6330	670
Total Croatia (moored at sea)	12,100	5613	5935	552
Split-Dalmatia County	2064	673	1178	213
Moored at sea	1697	439	1096	162
Land	367	234	82	51

Table 12 Number of vessels in transit in nautical tourism ports (2019) [21]

	Total	Motor yachts	Sailing boats	Other
Total Croatia	204,858	60,303	132,318	12,237
Total Croatia moored at sea	20,2412	58,892	131,373	12,147
Split-Dalmatia County	55,633	10,803	40,818	4012
Moored at sea	55,343	10,693	40,697	3,953
Land	290	110	121	59

2064 (673 motor boats, 1178 sailboats, and 213 others), with 82.2% of them being moored (Table 11).

Also, the number of recorded vessels in transit in Split-Dalmatia County in 2019 was 55,633, out of which 10,803 were motor vessels, 40,818 were sailing vessels and 4012 were other vessels. An overview of the number of vessels in transit in LNT on national and county level is shown in Table 12 [21].

The overview of marinas in Split-Dalmatia County is presented in Table 13, segmented by the criterion of the number of berths in the sea and on land. The data shows the spatial distribution of marinas, which also serve as a demographic and economic measure to increase activity in remote areas and contribute to sustainable development, especially for island destinations.

The data provided in Fig. 7 shows an almost uniform dynamics of the number of vessels and the number of days spent in nautical tourism ports on a permanent berth during the year. However, the representation of the number of vessels and the number of days spent in nautical tourism ports on a transit berth gradually increases with the beginning of the tourist season and culminates in period June–September, in which over 82% of total days of stay are realized.

If we hypothetically take into account the traffic of smaller ships, and indeed ships not equipped with a AIS device, the traffic at the entrance and in the immediate vicinity of the Port of Split increases significantly compared to the figures already shown. In the wider surroundings of the Port of Split, there are a large number of boats and other smaller vessels whose traffic significantly affects the overall safety of shipping in this area, especially during the summer months [9].

Additionally, it should be mentioned that the Port of Split was used as a seaplane port in the period from 2015 to 2016. The airstrip was located in the immediate

Table 13 Marinas in Split-Dalmatia County [22–25]

Marina	Position	Number of berths	Dry berths
ACI Marina Split	Split-City Port	364	60
Marina Kaštela	Kaštel Gomilica	420	200
ACI Marina Trogir	Trogir	174	35
Marina Zirona	Drvenik Veli	120	0
Marina Agana	Marina	134	70
ACI Marina Milna	Brač	170	15
Marina Vlaška Milna	Brač	71	0
ACI Marina Vrboska	Hvar	119	12
ACI Marina Palmižana	Hvar	180	0
Marina Tučepi	Tučepi	150	0
Marina Podgora	Podgora	220	0
Marina Baotić	Seget Donji	480	200
Lav	Podstrana	74	0
Baška Voda	Baška Voda	30	0
Marina Ramova	Krvavica	195	60
Marina Martinis Marchi	Maslinica	50	0
Marina Zlatan Otok	Hvar	40	0

vicinity of the port and crossed the paths of conventional shipping and boating routes. Significant traffic was generated in 2015 and 2016, but as of October 2016, this type of traffic has been suspended until further notice.

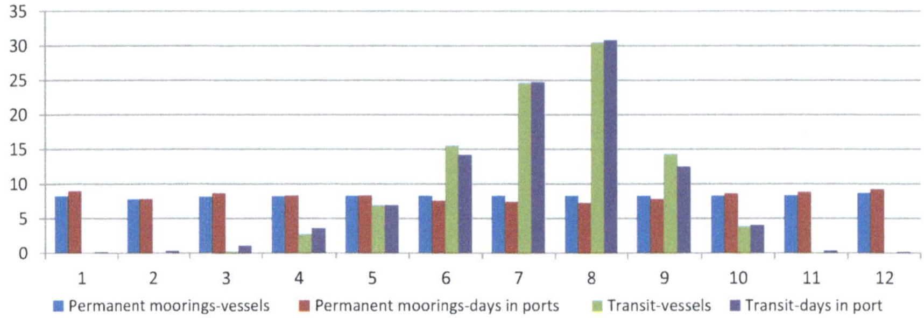

Fig. 7 Monthly dynamics (%) of the number of vessels and days of stay in nautical tourism ports (Croatia) in 2019 [21]

5 The Impact of the COVID-19 Pandemic and the Assessment of Future Traffic Using Linear Trend Projection

The overall port traffic in the Port of Split has generally been on an upward trend and the port is expected to continue to grow. The COVID-19 pandemic that started in 2020 has significantly affected the entire Croatian economy, including maritime transport, and it will certainly take several years to fully recover. Passenger and RO-RO traffic of the Port of Split is directly related to tourist traffic, which can be seen by comparing the decrease in this traffic activity and the decrease in the total traffic of the city port, both in the number of ship arrivals and in passenger and vehicle traffic. The number of ship arrivals in the Port of Split decreased by 44%, passenger traffic decreased by 62% and vehicle traffic decreased by 26% in 2020. The decrease is even greater in international scheduled traffic, where the traffic of passengers decreased by 91% and of vehicle units by 79%. At the same time, the number of tourist arrivals in Split, but also in the whole Split-Dalmatia County, has decreased by 65%. Moreover, the number of boaters in nautical tourism has also decreased by over 50%. In contrast, the number of registered ships and boats has not changed significantly in 2020, but only in the area of slightly higher activity in the depreciation process of ships, due to the obsolescence of boats for personal use. Table 14 shows tourist arrivals and overnight stays and tourism activity based on specific property types at the national level in 2019 and 2020.

Simultaneously with the overview of tourist activities at the regional and national level, the overview of the number of smaller vessels with home ports in Split and Kaštela is also given. The data presented in Fig. 8 show the sharp decrease in the total number of smaller vessels. In addition, a number of obsolete and older vessels were written off, a consequence that should be taken into account.

Until the pandemic COVID-19, the Port of Split recorded positive growth in all analysed categories of port traffic, with projected estimates of its continuation, especially in the segment of passenger and tourist ships and boats traffic. It is evident

Table 14 Tourist arrivals and overnight stays in 2020 based on the regional distribution and overall values in Croatia regarding the object type [26]

	1-XII 2020				1-XII 2019			
County	Domestic	Foreign	Total	% Arrivals	Domestic	Foreign	Total	% Arrivals
Split-Dalmatia	236,101	1,115,680	1,351,781	17.42	323,488	3,636,321	3,959,809	19.13
Total Dalmatia	713.546	2,592,391	3,305,937	42,59	948,248	8,503,355	9,451,603	45,65
Total Adriatic Sea	1.261.919	5,676,208	6,938,127	89,39	1.719.781	16,317,451	18,037,232	87,12
	1-XII 2020				1-XII 2019			
Object type	Arrivals	Nights	% Overnight stays		Arrivals	Nights	% Overnight stays	
Nautica	235,473	1,646,346	3.03		538,084	3,557,621	3.27%	
Total	7,761,717	54,394,810	100.00		20,703,683	108,672,888	100.00	

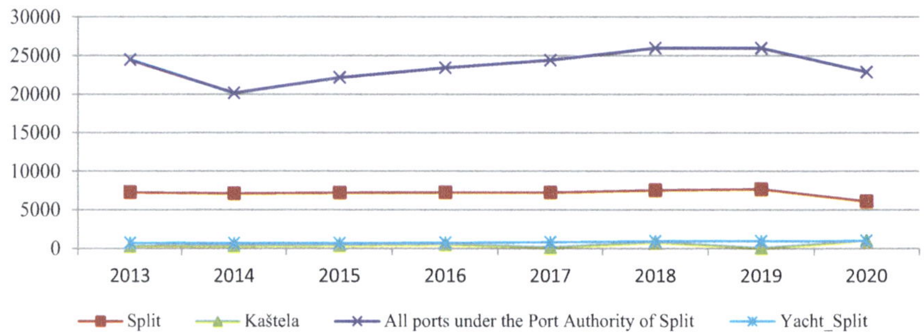

Fig. 8 Number of smaller vessels with port of registry in Split and Kaštela [12]

from Fig. 9 that the linear trend model describes the actual growth of ship traffic very well, until the crisis, with the trendline reliability (R2) of 0.81 for the City Port and 0.79 for the North Port. Figure 9 also shows a projection of the trend of recovery in vessel traffic, using the assumption of a five-year recovery for the City Port, which should correlate with the restoration of traffic to 2019 levels by 2024. This projection was performed using the Linear Trend Forecast option (Excel MS) based on known values of traffic up to 2020 and the assumption that in 2024 the traffic will reach 2019 figures (City Port). Year 2024 has been chosen because of the various analysis

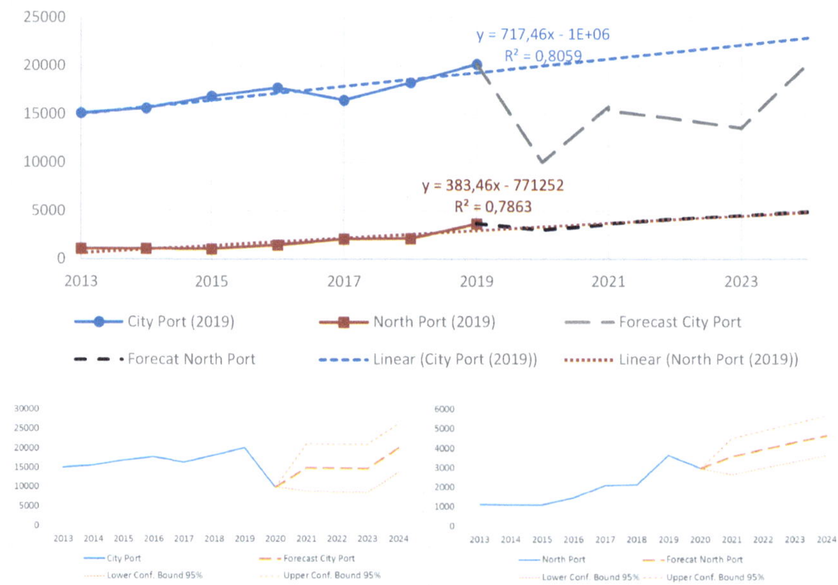

*Conf. Bounds (the lower part of the figure) - around the predicted value as a continuation of the timeline (constant step between the different points).

Fig. 9 Growth trend of port traffic (ship arrivals). *Source* Authors

(regarding GDP growth, transport sector, recovery of the tourism sector …) estimate that 2024 may bring an end of the crisis caused by the pandemic COVID-19 [27–31].

Linear regression equation is:

$$y = a + bx \tag{1}$$

where

$$a = \bar{y} - b\bar{x} \tag{2}$$

and b is:

$$b = \frac{\sum(x - \bar{x})(y - \bar{y}}{\sum(x - \overline{x})^2} \tag{3}$$

y's—known range of data (y values)

x's—known range of data (x values).

As already mentioned, presented trend in Fig. 9 is based on an estimate that the pandemic COVID-19 will end, or will almost end, by 2024. This pandemic is an unexpected, non-periodic, natural factor which is hard to predict, especially not by using the linear trend. However, as linear trend well described the growth of the Port of Split before the pandemic, it may be assumed that it could also predict the recovery of traffic, but with somewhat wider range of upper/lower confident (95%) interval, especially for the City Port. The positive growth forecast for the Port of Split, in general, is based on estimates of further growth of tourist activities in the Port of Split, but also on projects planned and under implementation, as follows:

- project of reconstruction and construction of new port infrastructure, i.e. construction of berths for receiving RO-RO vessels, the purpose of which is to shift domestic and international cargo traffic from the port basin of the city to the northern port, more precisely to the Stinice area,
- reconstruction of the eastern bank of the City Port,
- reconstruction of the tanker terminal in Solin (Kaštela basin C),
- reconstruction of road and railway infrastructure to increase the flow between the port and the hinterland,
- modernization of services in general, etc.

The impact of the pandemic COVID-19 in the first quarter of 2020 has reduced all predicted growth forecasts, as the global crisis has mainly affected the tourism sector and, accordingly, other activities that are directly or indirectly dependent on it. Port traffic in the Port of Split has decreased to the indicators recorded ten years ago. The decrease in the share of 44% in the arrival of ships is not even a true measure, considering that the domestic scheduled traffic contributes the largest part to the total port traffic and is still carried out with a minimal reduction in the number of ship departures, despite the pandemics. Cargo vessel traffic has also declined, but not nearly as much as passenger traffic. A much better indicator of the overall decline in port traffic at the Port of Split is the decline in passenger traffic, with limited cruise traffic and only small signs in the activity of international passenger lines.

6 Conclusion

The impact on a global scale, caused by the pandemic COVID-19, directly affected shipping traffic in the Port of Split. Although in the years preceding the pandemic, port traffic experienced continuous growth in all traffic segments (cargo, passengers, tourism, vessels arrivals …), statistical analysis of total port traffic shows that the impact of the COVID-19 pandemic drastically reduced the level of port traffic. If the criterion of the total number of vessels handled in the port is used to divide the port traffic into individual units, it is evident that domestic liner traffic still had positive indicators, but not to the same extent as in the pre-pandemic period. Domestic liner traffic refers to passenger and vehicle transport, which is necessary to maintain regular communication and supply of the islands and was maintained during the pandemic. The decrease in passenger and vehicle transport on these lines in 2020 is mainly related to the reduction of ship departures, especially during the peak tourist season. The activity of international liner traffic, which is closely related to tourist activity and depends on the arrival of tourists to the Croatian Adriatic coast, was kept at a minimum. The most significant decrease was recorded in passenger traffic on cruise ships. Compared to previous years, passenger traffic in this segment is almost non-existent. Cargo traffic in the Port of Split in 2020 recorded an increase, but number of vessels arrivals slightly decreased. The causal relationship between tourism and other activities on the example clearly proves that the global crisis has not only affected one segment of the economy, but has affected all others, especially the activity of passenger transport. Regardless of the temporary decline in activities in the port, further growth of the port of Split traffic is expected in the long run, especially the traffic of passengers and vehicles, i.e. the traffic of passenger ships. These positive trends have a foothold not only in the optimistic recovery of the tourism sector, but also in a whole series of already introduced projects related to the increase of port area and port capacities, improvements of the ports services, better connections with the hinterland, etc. All this should ultimately confirm and strengthen the status of the Port of Split as the largest and most important passenger port on the eastern Adriatic coast.

References

1. Mihanović V, Stanić D, Baljak K (2008) Specifičnosti luke Split zastupljenošću raznih vidova pomorskog prometa. Zbornik zajednice udruga inženjera Split. https://bib.irb.hr/datoteka/478635.Mihanovi_Stani_Baljak.pdf. Accessed: 20 Feb 2021
2. Split Port Authority-Docking Areas. http://portsplit.com/en/. Accessed: 20 Feb 2021
3. Dundović Č, Jugović A, Žgaljić D (2012) Analysis of Croatian ports in respect to Motorways of the Sea implementation. In: Proceedings of the 4th international maritime science conference, 2012 June 16–17, Split, Croatia
4. Vukić L, Lušić Z, Pušić D, Galić S (2019) Trends and perspectives of cargo traffic activities in the Port of Split. Naše More 2019—Conference proceedings
5. Maritime & Transport Business Solutions (MTBS) (2016) National Development Plan for Ports of Special International Economic Interest for Croatia, Ministry of the Sea, Transport and Infrastructure, Zagreb. Croatia. https://esavjetovanja.gov.hr/Documents/List/4553. Accessed: 14 Apr 2019
6. Vukić L, Ukić Boljat H, Slišković M (2018) Short Sea shipping—an opportunity for development of the North Port of Split. NAŠE MORE 65(3 Suppl):18–25. https://doi.org/10.17818/NM/2018/3.10
7. Split Port Authority (2020) Cargo & passenger and vehicle traffic 1990–2019, report
8. Split Port Authority (2021) Passenger and vehicle traffic 1990–2020, report
9. Marine traffic-Density maps (2018) https://www.marinetraffic.com/. Accessed: 18 May 2019
10. Lušić Z, Pušić D, Čorić M (2016) Maritime traffic on approach to port of split and assessment of collision and grounding risk. TOMS 5(2):130–140
11. Split Port Authority (2020) Report on the work of the Port Operations Center for 2019
12. Ministry of Sea (2020a) Transport and infrastructure. Data on Split port traffic
13. Harbour Masters Office (HMO) Split (2020) Report on work of the HMO Split for 2019
14. Ministry of Sea (2020b) Transport and infrastructure. The traffic structure by ship size for the City and North Port of Split
15. Marine Traffic (2019) Daily number of arrivals in the Port of Split—July/December 2019. https://www.marinetraffic.com/. Accessed: 18 Feb 2020
16. Jadrolinija-Sailing Schedule (2018) http://www.jadrolinija.hr. Accessed: 18 May 2019
17. Bura Line-Sailing Schedule (2018) http://buraline.com/plovni-red/. Accessed: 18 May 2019
18. Putovnica-Transport Overview (2018) http://www.putovnica.net/prijevoz. Accessed: 18 May 2019
19. Coastal Shipping Agency (2020) Traffic of passengers and vehicles on international lines in 2020. https://agencija-zolpp.hr/novosti/promet-putnika-i-vozila-na-medunarodnim-linijama-u-2020-godini/. Accessed: 18 Apr 2021
20. Croatian Bureau of Statistics (CBS) (2020a) FISHERY 2019—Provisional Data. https://www.dzs.hr/Hrv_Eng/publication/2020/01-04-01_01_2020.htm. Accessed: 18 Apr 2021
21. Croatian Bureau of Statistics (CBS) (2020b) Nautical tourism—capacity and turnover of ports 2019. https://www.dzs.hr/Hrv_Eng/publication/2019/04-03-04_01_2019.htm. Accessed: 18 Apr 2021
22. Marine u Hrvatskoj, CharterCroatia (2021) https://chartercroatia.net/hr/marine-hrvatska/. Accessed: 27 Apr 2021
23. Nautička Hrvatska (2019) https://www.htz.hr/sites/default/files/2019-12/TZ%20Marine%20HRVATSKI%202019%20za%20web.pdf. Accessed: 27 Apr 2021
24. Nautička Hrvatska, Sunlife (2021) https://www.sunlife.hr/hr/onama/marine-u-hrvatskoj.html. Accessed: 27 Apr 2021
25. ACI Marinas (2021) https://aci-marinas.com/. Accessed: 27 Apr 2021
26. Hrvatska turistička zajednica, Turistička statistika (2020) https://www.htz.hr/hr-HR/informacije-o-trzistima/informacije-o-tijeku-sezone/arhiva-2020. Accessed: 18 Apr 2021
27. Kissler1 SM, Tedijanto C, Goldstein E, Grad YH, Lipsitch M (2020) Projecting the transmission dynamics of SARS-CoV-2 through the postpandemic period. Science 368:860–868

28. EUROCONTROL (2020) Five-year forecast 2020–2024. https://www.eurocontrol.int/sites/default/files/2020-11/eurocontrol-five-year-forecast-europe-2020-2024.pdf. Accessed: 17 July 2021
29. Hermes E (2020) Tourism: Europe will be at the frontline of the recovery. https://www.eulerhermes.com/en_global/news-insights/economic-insights/Tourism-Europe-will-be-at-the-frontline-of-the-recovery-but-only-in-2024.html. Accessed: 17 July 2021
30. UNWTO (2021) International tourist arrivals, % change, 2021. https://www.unwto.org/taxonomy/term/347. Accessed: 21 July 2021
31. UNWTO (2021) COVID-19 and Tourism. https://www.unwto.org/covid-19-and-tourism-2020. Accessed: 21 July 2021

Assessing Barriers of Low-Carbon Emission Operations in Logistics Service Providers in Vietnam

Thi Thu Huong Tran, Van Kiem Pham, Phuong Thao Vu, Thu Trang Pham, and Khac Huy Nguyen

Abstract This paper assesses barriers to low-carbon emission operation in logistics services providers in Vietnam—a developing country which has witnessed the significant growth of logistics industry in the context of improving environmental sustainability. In-depth interviews and a survey with 159 logistics service firms were taken to collect data about eight groups of barriers to reduce carbon emission in practice (includes: cost, capability, complexity, information exchange, technology, partners, customers and government support). Among a variety of barriers, the most significant ones belong to cost, technology and government support groups. From understanding the extents and effects of each barrier, this study also points out some promising solutions which can be applied in Vietnamese logistics service companies.

Keywords Low-carbon emission · Barriers to low-carbon emissions · Logistics service provider · Vietnam

1 Introduction

Based on the Climate Change Knowledge Portal, Vietnam is one of the world's most vulnerable countries to climate change [1]. Having fully recognized the challenges posed by climate change, the Vietnamese authorities have issued many policies and programs to response to those difficulties and build a more sustainable, greener growth model is at the core of the development agenda accordingly. In particular, Vietnam ratified the 2016 Paris Agreement on Climate and committed to reducing greenhouse gas emissions by at least 8% by 2030 and to achieving the United Nations Sustainable Development Goals (SDGs) by 2030 [2]. The National Committee on Climate Change was also established and chaired by the Prime Minister and including key ministers to oversee climate change and green growth policies. These policies aim to better prepare Vietnam for the future impact of climate change should focus on lowering the intensity of fossil fuels in different sectors, investing in climate resilient

T. T. H. Tran · V. K. Pham · P. T. Vu (✉) · T. T. Pham · K. H. Nguyen
Department of Business Logistics, Thuongmai University, 79 Ho Tung Mau Street, Cau Giay District Hanoi, Vietnam
e-mail: thaovp@tmu.edu.vn

M. Petrović et al. (eds.), *The Science and Development of Transport—ZIRP 2021*,
https://doi.org/10.1007/978-3-030-97528-9_9

infrastructure or providing stronger incentives for firms to pursue green growth [2]. Among different sectors, logistics is one of the most significant ones which must be focused to reduce carbon emission due to its important role of main pollution sources and resource user at the same time [3]. However, low-carbon activities in logistics industry always must face with many barriers which inhibit their operations and efficiencies, especially in term of logistics services providers. In addition to provide traditional logistics management activities such as transportation, warehousing, order processing and so on, logistics services providers (LSPs) also evolve into a leadership role and have taken on a more proactive manner for customers by organizing networks, sharing information, managing assets and reducing inventory. LSPs are considered as both service provider and resource integrators which take responsibilities for a great volume of carbon emissions within logistics industry in accordance with the increasing logistics outsourcing [4].

In this article, starting from the investigation of recent low-carbon emission performances, we will assess the main barriers to reduce carbon emission in operation of logistics services providers in Vietnam, through which a framework to overcome these barriers may be developed. Motivated by the fact that logistics are an increasingly important contributor to carbon emissions in each country, the following research questions have been formulated:

RQ1. What is the level of low-carbon activities applied in logistics services providers in Vietnam recently?

RQ2. What are the greatest barriers that have inhibited the low-carbon emissions operation in logistics services providers in Vietnam?

RQ3. From the theoretical, policy and practical perspectives, how can the barriers to low-carbon operation in Vietnamese logistics providers be effectively mitigated?

LSPs, which mentioned in this research, refer to freight transportation, warehousing, distribution centre, forwarding, station and others. Separately, following conventional usage, "carbon emissions" refer to emissions of all GHG (not just carbon dioxide), with volume measured in tonnes of carbon dioxide-equivalents (CO_2e).

2 Literature Review

2.1 *Low-Carbon Emission*

Carbon emission can be understood as the release of carbon, or more specifically carbon dioxide (CO_2), into the atmosphere. Carbon dioxide is the main type of greenhouse gases, which contributes to creating global warming and some other environmental problems. Low-carbon emission means to reduce carbon emission to a low level to mitigate the effect of these problems.

Table 1 Some research papers made on low-carbon emission

Aspects of research	Papers
Driving factors of carbon emission	[5]
Impact of different policies on reducing carbon emission	[4, 6–9, 33]
Influencing factors to low-carbon emission	[10–13]
Barriers against low-carbon emission operations	[14–18, 20–24]
Drivers for low-carbon operations	[17, 25, 26]
Model simulation for low-carbon emission	[27, 28]
Review of technology development	[29]
Impact of low-carbon emission on business performance	[30]

Source Summarized by authors

Due to its crucial role in mitigating global warming situation, low-carbon emission has become a topic of interest in research recently. Regarding low-carbon emission within the economy, researches have been done on various aspects such as: driving factors of carbon emission [5], impact of different policies on reducing carbon emission [4, 6–9], influencing factors to low-carbon emission [10–13], barriers against low-carbon emission operations [14–24], drivers for low-carbon operations [17, 25, 26], model simulation for low-carbon emission [27, 28], review of technology development [29] and impact of low-carbon emission on business performance [30].

The majority of those studies focuses on aspects such as impact of different policies on reducing carbon emission, influencing factors to low-carbon emission, barriers against low-carbon emission operations and drivers for low-carbon operations. Besides, some other aspects have also been considered, such as household low-carbon emission [31] and low-carbon emission diets [32] (Table 1).

In general, there is a considerable number of researches about different aspects of low-carbon emission in various countries such as the UK, Denmark, Germany, the USA, Australia, Greece, Italy and China. However, there is not any research regarding barriers against low-carbon emission operations in Vietnam available.

2.2 Low-Carbon Emission in Logistics

Logistics and supply chain management has now proved their vital role in businesses and economy. Sustainable development of logistics and supply chain, thus, has become an indispensable decision for all logistics and supply chain management. Among all the initiatives for a sustainable logistics, green logistics has emerged as a goal for all companies: manufacturers, trading companies, logistics service providers… and even end customers; and low-carbon emission is one of the factors to be considered to achieve a green logistics activity. In this part of the paper, the group of authors will review some key points concluded from some recent research on this matter, focusing on low-carbon emission applications on logistics activities and services.

Current available research has not studied specific drivers for low-carbon emission application in logistics activities and services. However, it is possible to conclude some probable drivers for low-carbon emission from related topics such as: green logistics or low-carbon operations in supply chain. Lin and Ho [22] suggested that green practice's relative advantage, green practice's compatibility, organizational support, the quality of human resources, customer pressure, regulatory pressure and governmental support could motivate the application of green logistics in China logistics companies. Some of the above-mentioned drivers are agreed by Evangelista [17] in his research on green logistics of some companies in Italy. Specifically, according to Evangelista [17], green logistics adoption of some companies could be driven from actions of their competitors, customers, support from government and associations. Since low-carbon emission is a part of green logistics, it is appropriate to deduct from these two research papers that some drivers for green logistics could also be drivers for low-carbon emission, such as: pressure from competitors, customers and governments, relative advantage, support from governments and highly capable human resources. On the other hand, some of these drivers are also mentioned in some research on low-carbon operations in supply chain. Jabbour et al. [26] summarized some drivers for organizational low-carbon actions, for example, profit, ethical considerations, cost savings, regulatory compliance, competitive advantage, regulation, requirement of customer and company green image. Some of these drivers such as company green image, regulatory requirement, competitive pressure, government regulation or requirements from customers have also been mentioned in the research of Ibrahim et al. [25]. It can be clearly seen that most of the drivers for low-carbon actions in supply chain perspective are similar to the ones in green logistics aspect. Some other drivers are mentioned including a green image of company and profit has been suggested.

Even though low-carbon emission is a part of green logistics and has become increasingly aware of by companies, the impact of it on business performance of a company is limited [30]. However, it is worth to note that the research of Trivellas et al. [30] focuses on the impact of logistics emission on companies other than logistics service providers. Thus, the result from this research cannot deny the great impact of low-carbon emission on logistics service provider performance and the development of a sustainable logistics and supply chain, and further research on this matter is necessary.

2.3 Barriers to Apply Low-Carbon Emission Practices in Logistics

The number of literatures on carbon footprint in logistics sectors is quite limited and mostly approaches the research issues from the perspective of supply chain management. As discussed above, low-carbon emission can be considered as a specific segment in green logistics initiatives, belonging to the field of practices for sustain able logistics. For this reason, a number of studies directly researching or discussing obstacles to implementation of sustainable or green activities of logistics providers have been reviewed, revealing eight potential categories of barriers as below.

Cost

Cost has been generally concerned as an obstacle to the adoption of practices contributing to green initiative, particularly carbon deduction. Specifically, investment is what worries logistics providers [17]. In order to conduct the practices, extra investment is needed, and the fact that possibility of payback is unguaranteed [21] may discourage LSPs to implement.

Capability

Ability to conduct the measures is another constraint, which can be elaborated as technical understanding and skills. Since sustainability or green initiatives are recently put into consideration recently, some logistics companies, particularly small and medium ones, may lack of knowledge specializing in green solutions and techniques needed to manage them [20]. This shortage can be also interpreted in the shortage of human resources to work on those practices [17].

In addition, this factor can also be constructed by the lack of sufficient and proper education and training [24]. Employees and managers may be well aware of the necessity and research about solutions to reduce carbon releases but are unsure or confused about how to implement properly, which impede firms from prompt adoption.

Complexity

As explained above, these practices require technical knowledge and understanding and are normally added into current working process of logistics companies. Hence, complexity could take place in the introductory phase within forces in organizations. Certainly, implementing phase could trigger complexity of coordinating different divisions and still remaining efficient performance [22]. In some emerging markets, sustainable initiatives are not ubiquitous [23] particularly in reducing carbon emission; therefore, complicated measurement to track carbon releases from operations is also an obstacle.

Information Exchange

For some logistics providers, they suppose skills of communicating and exchanging information when practising these initiatives as one of the top elements that may hinder the implementation [17]. Precisely, the information could be transferred insufficiently or even missed between different levels or among departments, owning to the fact that communication forms, such as meetings, explanation or consultations, are omitted [15]. Specific types of information could be addressed as company regulations, policies or guidelines [17], objectives of the implementation and outcomes and impacts of the practice application [19].

Technology

Application of the practices require technologies to facilitate which could be unavailable in some markets [20]. In some situations, companies are worried if those technologies cannot fit in the current system and they may have difficulties in integrating them into their operation processes [16].

Partners

In the process of completing logistics services to customers, companies may outsource some activities to partners. This outsourcing brings about an external barrier when those contracted partners are incompetent in following the green practices. The incompetency could be interpreted as limited financial resources or lack of labour forces to conduct [18]. Govindan et al. [20] also added that the outsourced service providers could be unwilling to follow the request of conducting environmental practices when some operational information and techniques have to be exchanged.

Clients

Another external source of obstacle is the clients when they are unaware of the necessities of sustainable implementations [17]. Even if they are aware, those practices are not operated within their business; therefore, what they mind is time to fulfil the task and charges on the services [14]. The unwillingness and different priorities from customer perspectives are potentially considerable barriers for logistics firms in adopting carbon reduction techniques.

Government

As a contemporary issue, sustainable initiatives in general have been not yet clearly defined or elaborated in governmental document. The process of implementation is hindered when logistics firms have inadequate motives owning to the lack of official legal framework, for instance, regulations [14, 17]. Additionally, there is a possibility that the current legal documents may conflict with regulations to adopt sustainable practices [34], which confuses firms and discourages them in implementing.

3 Methodology and Data Collection

Both qualitative and quantitative approaches were applied in this research. For the first phase, the author looked into a range of secondary data (i.e. statistics and information from international and national published studies, published reports and other types of publications by Vietnamese Ministry of Trade and Industry, Vietnamese Ministry of Natural Resources and Environment, General Statistics Office of Vietnam) to review and systematize current status of practices and obstacles to low-carbon emission operation in logistics service providers. Based on the results, 09 semi-structured interviews, which included a number of low-carbon initiative and barriers reviewed from literatures, were conducted. These 09 respondents are senior managers or above of prominent logistics service providers in Vietnam, both domestic and foreign invested ones, whose profiles are summarized anonymously in Appendix A. The content of the interviews focused on viewpoints of management about current practices and adaptability with current barriers and requirements from Vietnamese businesses and market.

In the second phase, revised surveys were massively distributed to 279 logistics service providers via various means (i.e. in-person, Google form). The authors managed to contact these LSPs via the network of Vietnam Logistics Association (VLA)—a professional organization of Vietnamese organizations, businesses and citizens in providing logistics services listed under the Commercial Law. In return, the authors received 166 answers accounting for 59.5%—a relatively high return rate [35], of which 159 were eligible for in-depth calculation and analysis.

All 159 respondents are carrying out some of key logistics services such as transportation (air, sea and road), logistics part (warehousing, distribution). Over 50% have experienced in logistics industry in Vietnam for more than five years, majorly operated in the two major Vietnam logistics hubs—Hai Phong and Hanoi in the North and Hochiminh City and Vung Tau in the South. This corresponds with the locations of two biggest ports of this country: Hai Phong and Hochiminh City (Sample details in Appendix B). For this reason, the sample should be credible enough for further analysis and implication generation.

SPSS 20 software was utilized for processing the data for the quantitative approach. Answers were collected on the scale of 5, and then the group of authors performed some analysis such as: calculating mean and standard deviation, analysing statistically and thoroughly figures collected on current measures of low-carbon emission in transportation, warehouse, packaging… and analysing barriers preventing businesses from applying low-carbon emission measures. Detailed results of these analysis will be discussed in the next section of this paper.

4 Results and Discussion

4.1 The Level of Low-Carbon Operation Applied in Logistics Services Providers in Vietnam

Before assessing the barriers in reducing carbon emissions in Vietnamese Logistics services providers, invitations to the survey described above were sent to the targeted respondents to firstly investigate the level of actual carbon abatement activities applied in each respondent. These activities are classified in four group: freight transportation, warehousing, packaging and information system.

Freight transportation is significant energy consumer-related greenhouse gas emission [36]. Therefore, cutting carbon emission in freight transportation is always considered as one of the most important activity in green logistics. As illustrated in Fig. 1, the greatest activity applied in freight transportation in logistics services providers to reduce the carbon emission is minimizing the journey without goods. On the one hand, companies can be proactive in applying that activity and, on the other hand, its impact immediately to the cost of each company as a key motivation to keep continuing that operation. There are nearly 72% of respondents said that they "usually" and "always" applied minimizing the "zero" truckload. The second most used solution to lower the greenhouse gas emission is using environmentally friendly fuels; instead of fossil fuel, 47% of companies responded that they use biofuel regularly. Other operations like installing emission control system for vehicles are difficult to use as the investment required is costly, and the environmental inspection in Vietnam has not been strictly implemented. Transition from road to railways and water ways transport also struggles to be implemented because the nature of carrying good leads to be transferred by road. The difference between the applying

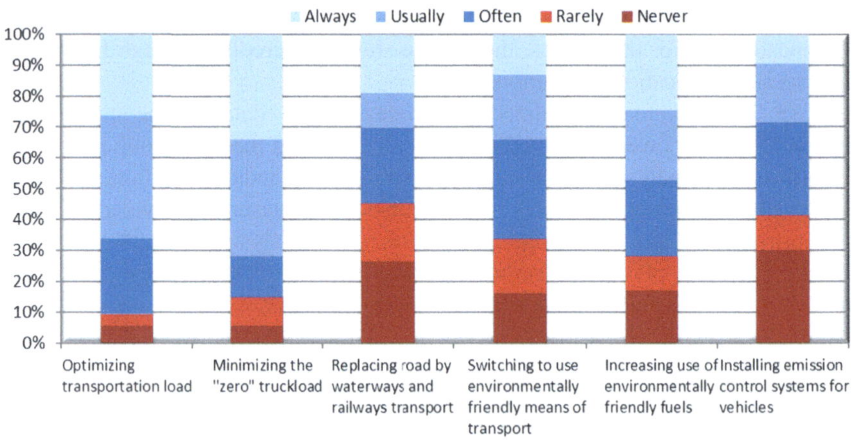

Fig. 1 Frequency of low-carbon operation applied in freight transportation. *Source* Collected by authors

frequencies of each operation can be explained by the diverse in size, services list and platform of respondents.

In warehousing operation, it is clear that the most popular action applied to cut down carbon emission is setting warehouses at convenient traffic areas, with 75.5% companies use that option more than often. LSPs believe that the advantages of warehouse location will help to save time and energy using in transportation; then the carbon emission is decreased accordingly. The other activities like using energy saving equipment or using recycled materials in building warehouse… have similar applied level as "often" applied. The lowest volume activity is using solar energy system with average index of frequency equals 2.4 (Fig. 2).

As can be illustrated in Fig. 3, data about low-carbon operation applied in packaging is consistent. As be seen in Fig. 3, almost average is higher than 3; it means that almost asked companies use carbon abatement solutions in packaging but not regularly. Among six solutions in related to packaging, optimizing the amount of goods contained in each packaging and managing used packaging are the most used activities, around 70% of asked logistics services providers operated those activities as usually and more.

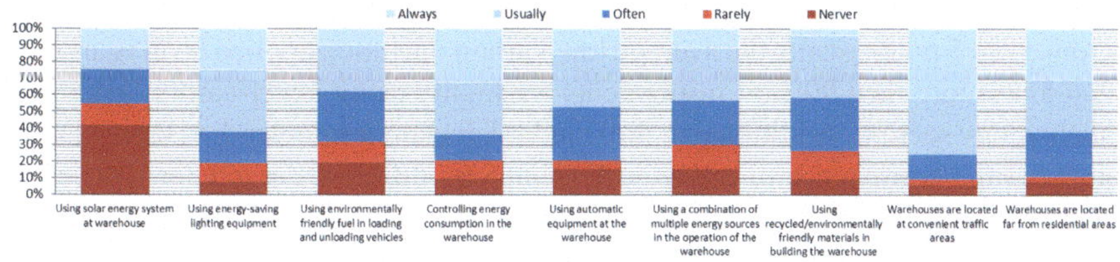

Fig. 2 Frequency of low-carbon operation applied in warehousing system. *Source* Collected by authors

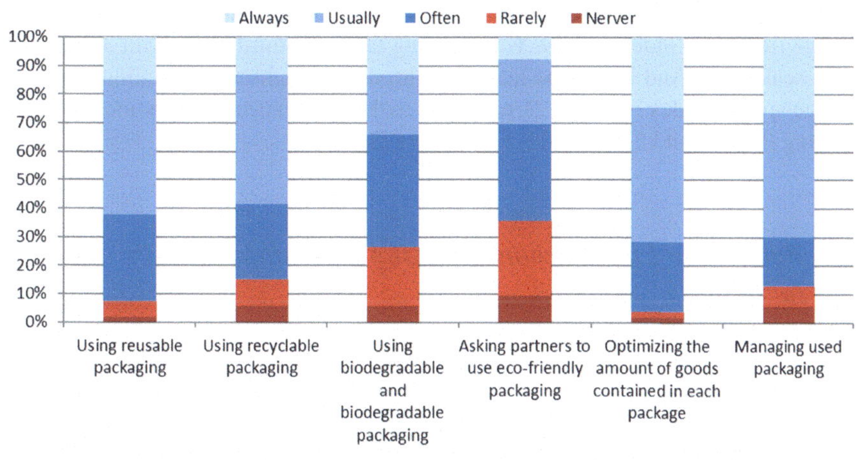

Fig. 3 Frequency of low-carbon operation applied in packaging. *Source* Collected by authors

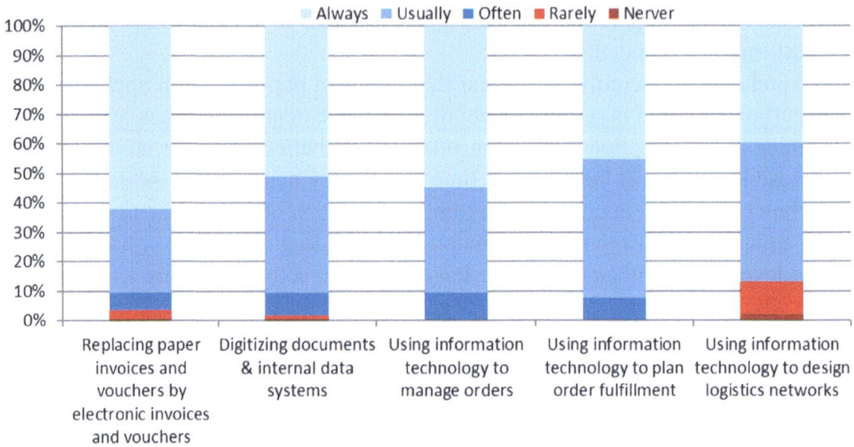

Fig. 4 The frequency of low-carbon operation applied in technology. *Source* Collected by authors

Comparing with the three other groups, responses from Vietnamese LSPs show that they apply low-carbon emission operation in technology much more than any sectors. On a 5 points scale, the average low-carbon operation applied level is around 4.2–4.5, as more than 90% of respondents are highly using lower-carbon solutions in technology section. Additionally, the standard deviation is lower than 0.8 and also points out the consistence of applying level among different companies. It is easy to understand as technology is a vehicle to enhance the overall effectiveness and efficiency of logistics system, which not only help to simplify the task but also save time, cost and laborious needed in operation [37].

Overall, the survey shows that decreasing carbon emission operations is most applied within technology system group as investing in technology is the development trend of logistics industry. The second most applied one which is packaging with average frequency index is 3.49. Warehousing takes the third place with 3.31 point in the scale of 5. And lastly, the least used group is freight transportation, and the average applied index is only 3.21 point due to the high expenses investment cost of applying (Figs. 4 and 5).

4.2 *Barriers to Low-Carbon Emission Operation in LSPs in Vietnam*

In Table 2, there are 14 out of 23 criteria having their means within the range from 3.00 to 3.40, namely: CO3, CA1, CA2, CA3, CM1, CM2, CM3, IE1, IE2, IE3, PA1, PA2, CL2, GO4. It can be referred that even though averagely respondents gave a positive judging towards these factors influence on low-carbon initiatives, their levels of agreement on these criteria are relatively more neutral than on the

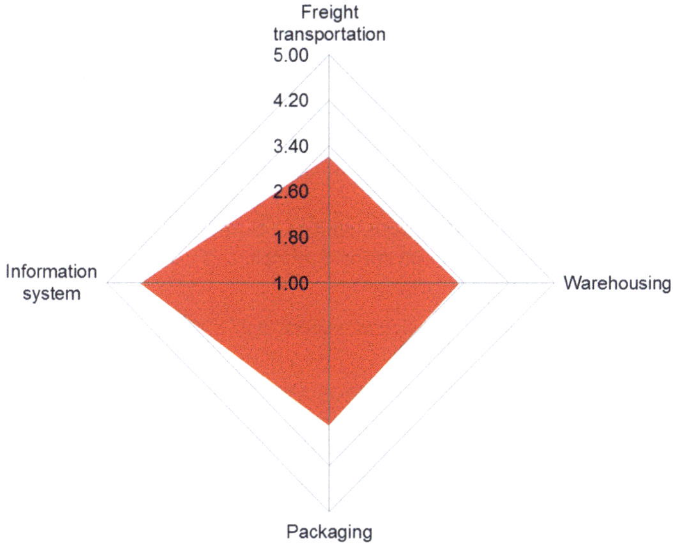

Fig. 5 Mean of adoption level of low-carbon emissions of surveyed LSPs. *Source* Collected by authors

others. Among those, complexity, information exchange and partners are the three categories having all constructing criteria in this group. Hence, from perspective of LSPs operating in Vietnam, these three segments are not significantly troubling them to following the green movement. Particularly, these three barrier groups are very demanding in terms of interaction between various stakeholders in the logistics service chain. These findings reveal the fact that LSPs in Vietnam are not seriously concerned about calling for cooperation between different parties or different division within their firms to join hand in adopting low-carbon reduction.

The other nine criteria, including: CO1, CO2, CA4, TE1, TE2, CL1, GO1, GO2, GO3, have the mean scores falling in the range between 3.41 and 3.85, implying a positive agreement that these are obstacles to their low-carbon emission operations. Among those, cost and government are the two categories raising much concerned among surveyed LSPs. Each has two criteria achieving top average scores. In addition to expenses and governmental factors, the services providers suppose customer priority in outsourcing services and technological integration to be worrying fencing (Fig. 6).

As being revealed from Table 2, the average scores of all 23 criteria are from 3.00 and above. Criteria CO2 about investment cost account for the highest mean score of 3.85, while the lowest was 3.00, belonging to a criteria of information exchange—IE2 (objectives of low-carbon emission practices are not transmitted well within the company). Additionally, it is well noted that results of all criteria have small standard deviations which are all less than their corresponding mean score. The more than moderate means together with small standard deviation may imply that all surveyed

Table 2 Results of surveyed LSPs about barriers to low-carbon emission operations

		Mean	Standard deviation
Cost			
CO1	Cost of using low-carbon emission energy is high	3.66	0.973
CO2	Cost of investment in low-carbon emission facilities is high	3.85	0.901
CO3	Cost of educating and training human resources to implement low-carbon emission practices is high	3.18	0.800
Capability			
CA1	Understanding of carbon emission measurement is limited	3.25	0.953
CA2	Understanding of low-carbon emission practices is limited	3.25	0.933
CA3	Updating knowledge in low-carbon emissions is limited	3.30	0.986
CA4	There is a lack of training programs on low-carbon emission for human resources	3.45	1.077
Complexity			
CM1	It is complex to introduce low-carbon emission practices in the company	3.08	0.952
CM2	The implementation of low-carbon emission practices is complicated	3.26	0.938
CM3	Measurement of carbon footprint is complicated	3.32	0.930
Information exchange			
IE1	Company's policies on low-carbon emission practices are not delivered clearly	3.04	1.067
IE2	Objectives of low-carbon emission practices are not transmitted well within the company	3.00	1.049
IE3	Outcomes and impacts of low-carbon emission practices is not transmitted well within the company	3.02	1.058
Technology			
TE1	Accessibility to new technologies of low-carbon emission is limited	3.53	1.042
TE2	Integration of low-carbon emission technologies into the current system is difficult	3.58	1.039
Partners			
PA1	Partners are incompetent to cooperate in conducting low-carbon emission practices	3.32	0.970
PA2	Partners are uncooperative in conducting low-carbon emissions practices	3.15	1.020
Clients			

(continued)

Table 2 (continued)

		Mean	Standard deviation
CL1	Customers prioritize other benefits over low-carbon emission practices	3.75	1.151
CL2	Customers lack awareness of low-carbon emission practices	3.15	0.966
Government			
GO1	There is a lack of government guidelines on low-carbon emissions	3.58	1.110
GO2	There is a lack of national standards for low-carbon emissions	3.49	1.024
GO3	There is a lack of government incentives and support for low-carbon emissions	3.70	1.078
GO4	Some current government regulations hinder the implementation of low-carbon emission practices	3.28	1.020

Note 1 = Strongly disagree; 5 = Strongly agree
Source Collected by authors

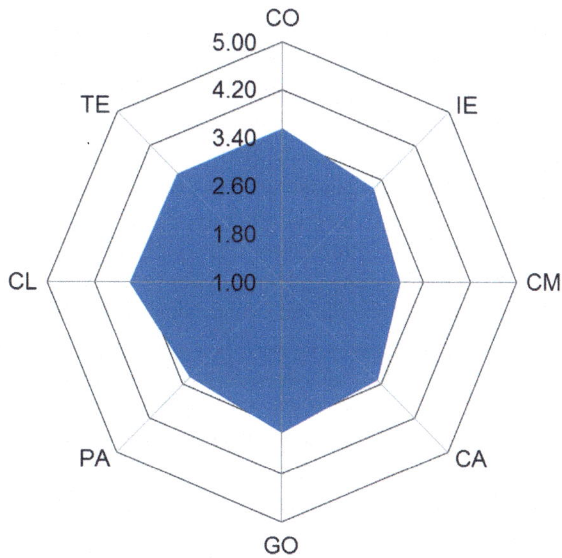

Fig. 6 Mean of eight groups of barriers to low-carbon emission practices. *Source* Collected by authors

LSPs, to some extent, agree the proposed criteria as barriers to their implementation of low-carbon emission reduction.

Problems with cost of investment and low-carbon emission energy

The criteria CO2—cost of investment in low-carbon emission facilities holds the highest score of 3.85 among all 23 criteria. Precisely, 73.6% of respondents agreed that this hindered their operations to cut carbon releases, among which more than a fifth of all surveyed LSPs totally agree. Another cost factor that is CO1—cost of using low-carbon emission energy is highly ranked as the fourth highest average result with 3.66 and also got a fifth of the respondents strongly supporting as a barrier.

LSPs' concerns about financial issues may root from the scale of logistics companies in Vietnam. According to a report of Orbis research, currently, 3000 companies offering logistics services in Vietnam, with a majority (90%) having less-than-VND 10 billion register capital. 5% have their registered capital within VND 10–20 billion, and just 5% have more than VND 20 billion registered capital [38]. In the meanwhile, a number of low-carbon initiatives require investment into new facilities and infrastructure. Firstly, facilities supporting low-carbon emission make include solar panels, electric vehicles and cranes, building renovated to achieve LEED certification and so on. For example, a common low-carbon emission implementation conducted at ports is using crane (e.g. quay cranes, RTG cranes, etc.) powered by electric instead of diesel. In such cases, a few of large-scale port operators are capable of investing in purchasing the new ones while considering whether the extra productivity is needed in comparison with current customer demand or not. For large-scale logistics firms in Vietnam, initial investment may not worry them very much, particularly with some adoptions like solar energy installation having approximately five to seven-year depreciation period. However, they still emphasize a number of incurring investment costs of replacing expired solar batteries, purchasing back-up panels, dissolving unusable power banks. Secondly, some low-carbon initiatives may demand extra infrastructure to be implemented in Vietnam. For instance, some freight businesses would like switch to greener transportation modes, i.e. from air to intercontinental rail from Vietnam to Europe or from truck to inland waterway barge for domestic shipment. However, in order to ensure transport efficiency when delivering by barges, there should be a network of modern and well-equipped ICDs and depots connecting with sea terminals, which have been neglected, especially in the Northern Vietnam.

Shifting to low-carbon energy is remarkably worried as a cost barrier. As can be extracted from the status analysis, renewable energy usage is not common among surveyed LSPs, which probably has a link with their concern of investment (CO2) into facilities (e.g. solar panels) discussed above. Besides renewable power, some greener alternatives like E5 fuel, E10 fuel, CNG and so on gradually entered Vietnam market. However, application of them is still limited for a number of reasons. Being introduced in Vietnam in 2007, E5 fuel is not favoured among transport businesses, particularly truck haulers, owning to their costly price—approximately 25% higher than diesel oil (PV oil). E10 with higher level of ethanol and believed to be even greener than E5 has yet produced and supplied widely. By contrast, compressed natural gas (CNG) with 84% formed by methane is nearly a perfect option whose cost is 30–40% less than FO, DO and LPG in operations. What makes CNG still a risky choice in terms of cost is the limited number of suppliers. Apart from Southern Gas Trading JSC, which currently supplies for some bus operator in the Hochiminh city, Pvgazprom natural gas for vehicles limited liability company has had planned to supply after 2020. However, the former recently had to cut supplies to transport partners by 20–30% in order to offer for electric production as requirement of Ministry of Industry and Trade [39].

Lack of government support as an obstacle

The criteria GO3—there is a lack of government incentives and support for low-carbon emissions holds the third highest score of 3.70, with more than a quarter strongly supporting as a barrier. Following GO3 in governmental barrier category, GO1—there is a lack of government guidelines on low-carbon emissions got a noticeably high average score of 3.58, with 24.5% of respondents strongly voting as an obstacle to their green practices.

Indeed, governmental support, particularly monetary one, is quite limited. For instance, the case of biofuel (E5, E10), although government offered preferential plans on land, finance and corporate tax for biofuel stake holders but none on selling price. Elaborating this decision, Ministry of Industry and Trading explained that it was impossible to have any distinction between domestic or foreign enterprises, state-owned or private enterprises and all are given the same privileges. Depending on business conditions, each enterprise can self-reduce prices or not to create initial incentives for E5 fuel consumers. The state will not compensate prices and bear losses on behalf of petroleum enterprises.

Furthermore, logistics enterprises felt confused and unmotivated to apply practices since the unavailability of any legal document on low-carbon emissions of logistics, or even green/sustainable logistics in general. Actually, the authority did form a small number of circular or decision, but just within a specific logistics aspect, for instance, a mode of transport. To be more exact, circular 48/2015/TT-BGTVT [40], added and amended by circular 26/2020/TT-BGTVT [41] include regulations on registration of inland waterway vessel. Within the scope of the circular, it only requires the vessel not to discharge contaminating sewage into the environment. Decision 1456/QD-BGTVT [42] on action plan to respond to climate change and green growth for the period 2016–2020 did mention two objectives in related to low-carbon emission practices: to restructure transport market share, raising the usage of train and inland waterway transportation, and to promote the usage of renewable and green energy. Nevertheless, for the former, this decision not yet specify measure or performance indicator. For the latter, there was a target, but set for bus and taxi business, while for commodity transport commerce, it was only suggested to encourage service provider to change fuel type. There is another legal document specifically on carbon emission, which is Decision 4206/QD-BGTVT [43] on action plan to reduce CO2 emissions in Vietnam's civil aviation activities in the 2016–2020 period. However, the agenda of this decision is irrelevant with the subject of this research while solely about civil service.

The issue of customers' benefit priorities

The criteria CL1—clients prioritize other benefits over low-carbon emission practices was the second highest mean of 3.75 among all 23 criteria, especially more than a third of respondents strongly believed CL1 to be a barrier to low-carbon implementation. It is necessary to consider this judgement within Vietnam logistics context. According to World Bank report, logistics cost in Vietnam is significantly higher than in other ASEAN nations like Thailand or Singapore. Logistics cost, including transport, warehousing, customs clearance, in Vietnam, is equivalent to approximately 20.9–25% of GDP, while the global average is 12%. Specifically, transport cost in Vietnam is 6% higher than in Thailand, 12% higher than in Malaysia and triple the same type of cost in Singapore [44].

For this reason, shippers or logistics service customers in general are more cost sensitive when they outsource logistics service in Vietnam, which may affect the competitive advantages. Meanwhile, from what has been discussed above, initiatives to deduce carbon emission have tendency to raise cost. For example, on the progress of switching to cleaner fuel, shipping liners will adjust Bunker Adjustment Factor (BAF), then enhancing surcharge to $100/TEU [45], in some cases accounting for 10–15% of sea freight. In the long run, logistics companies, particularly forwarders, have to share the cost raise with their customers if they would like to sustain their business. If so, the impact on customer satisfaction and loyalty is the main worry.

5 Conclusion

It can be summarized from the analysis that logistics service providers in Vietnam are currently aware of the necessity of low-carbon emission operations and are applying various measures to reduce carbon emission to some certain extent. Some of the most popular measures in Vietnamese logistics service providers are minimizing the "zero" truckload, optimizing transportation load, locating warehouse at convenient areas, using recyclable and reusable packaging. However, there are a number of barriers that are hindering low-carbon emission operations in Vietnamese logistics service providers, divided into eight groups: cost, information exchange, complexity, knowledge, government, partner, customer and technology. According to the survey results, Vietnamese 2PL and 3PL service providers are facing the biggest challenge from cost, government and customers. The initial investment cost for low-carbon emission is still high, while the return has not been proved to be significantly favourable, which leads to the reluctance in adopting low-carbon operations. Meanwhile, the government has not been providing enough support to promote low-carbon emission application in businesses. Another large barrier is that customers have been reluctant to choose services with lower-carbon emission but higher cost; in other words, customers still value cost over low-carbon emission and refuse to make the trade-off in short term.

In order to mitigate the effects from such barriers and promote the application of low-carbon emission, we would suggest some solutions to both government and logistics service providers. Firstly, regarding government, they should issue some official guidelines to instruct businesses as well as set a standard or benchmark for consistent application in all businesses. This should help remove some barriers relating to knowledge or technology and help encourage businesses to reduce their carbon emission. The government should also offer some types of financial incentives to businesses such as applying lower tax rates to low-carbon emission equipment and vehicles import, offering reasonable financial aid to businesses willing to start to apply low-carbon emission operations. Besides, the government could run some campaigns to raise awareness of all businesses about the necessity and importance of low-carbon and its benefits to the businesses in the long term. This could to some extent encourage logistics service providers to start or extend their low-carbon activities as well as their customers to opt for services with low-carbon emission instead of choosing the cheapest service option. Secondly, regarding logistics service providers, they should cooperate with each other to share some facilities and equipment to optimize activities, leading to lower-carbon emission. In addition, they should cooperate also to share knowledge and experience in applying low-carbon emission initiatives, so that the barriers relating to knowledge, cost, complexity and technology can be mitigated. Logistics service providers should also proactively look for support from the government, share their thoughts and raise their ideas so that the government can support them better and gradually remove the barrier relating to government and information exchange.

The results of this research provide some insights on current barriers that Vietnamese logistics are facing in their application of low-carbon emission initiatives. These are supposed to support government and logistics service providers in coming up with solutions to encourage and facilitate further low-carbon emission operations. Nevertheless, this paper has managed to cover a number of specific sectors which may need further in-depth studies in the future, for instance, correlating relationship between barriers and decision on application or implementating efficiency of the initiatives. The extension of barrier impacts on each types of businesses according to ownership could be valuable to look into. Further research on different aspects regarding barriers to low-carbon emission operation in logistics service providers in Vietnam could provide more insights and help facilitate long-term and wide-spread application of it.

Appendix A: Profiles of Companies and Interviewees

No.	Business profile	Interviewee's position
1	A logistics services provider which offers multimodal transport services include: road, inland waterway and seaway Operating and exploiting the chain of multilogistics services for transporting container goods by barge through the system of inland waterway ports in the North of Vietnam	General Director

(continued)

(continued)

No.	Business profile	Interviewee's position
2	One of the leading shipping companies in Vietnam which acts as the general agent for Evergreen Marine Corp. with 1 head office (Ho Chi Minh city), 3 branches (Hanoi city, Hai Phong city and Da Nang city) and 1 sub-agent (Qui Nhon city) Globally, this company has the 4th largest container fleet in the world, with 180 boards provides a capacity of approximately 650,000 TEUs. In addition, having more than 240 agents and offices, the company is available to operate in over 80 countries with the shipping lines crossing East West–Southeast Asia, Hong Kong, Taiwan, Mainland China, Korea and Japan, the East and West coasts of the USA	Executive
3	A logistics services company which provides international freight forwarding services by seaway, airway and multimodal transportation. In which, ICD Port Service includes inland clearance point; container yards for imports and exports; container transportation and transshipment services by inland waterways; Warehouse Services (CFS warehouses, bonded warehouses, cold storages, general warehouses and distribution warehouses); Trucking Services (LTL & FTL) In 2020, this company invested in solar power system by setting a special roof at an ICD Port with a total area of 6600 m^2, capacity of 718 kWp and another roof with an area of 8600 m^2, capacity of 916 kWp. The total capacity of the two rooftop solar energy projects is 1634 kWp, the total average annual output will nearly 2333 Mwh	CEO
4	Transport international import/export goods by sea (FCL/LCL) CY-CY, Door to Door; International Export/Import goods by air, Door to Door service; Packing service for export goods; Warehousing services… Key customers: Proter & Gamble; Unilever; SC Johnson; Henkel	Manager
5	A services provider of receiving, packing and wrapping; transporting retail goods, whole containers and all kinds of goods; port handling and inland transportation; services of contractors transporting retail and dangerous goods	Vice director
6	This company changed to operate under the model of Joint Stock Company with key activities like: loading and unloading goods; transportation support services: ship brokerage, shipping agency, shipping agency…; towing and ship support services; warehousing and storage of goods…	Vice director
7	Operating in the field of logistics and freight on the main route connecting China–Vietnam–Laos/Thailand/Cambodia. The warehousing system and means of transportation are invested heavily	Director
8	As the 14th company of a France's Logistic Group, this company was established in June 2017 specializing in logistics services	Operation Manager

(continued)

(continued)

No.	Business profile	Interviewee's position
9	One of the leading shipping liners, having headquarter in Europe and mainly based in Hochiminh city. This firm has branches in Hanoi and across main ports in Vietnam. In addition to international container shipping, the firm is expanding in land-based logistics activities such as warehousing, intermodal and customs clearance. This firm is also an industry leader and pioneer in sustainability and digitalization	Manager

Appendix B: Demographics of LSPs Participating in the Survey (159 Responses)

Type of ownership	Frequency	Percent
Company with more than 50% state-owned stake	6	3.8
Company with 100% capital from foreign investors	27	17.0
Joint Stock Company	60	37.7
Private company	9	5.7
Joint venture with foreign investors	6	3.8
Limited liability	48	30.2
Others	3	1.9
Type of logistics services		
Road	120	75.47
Sea	93	58.49
Railway	42	26.42
Air	81	50.94
Logistics park	90	56.60
Parcel delivery	27	16.98
Seaport	45	28.30
Railway station	3	1.89
Airport	9	5.66
Others	30	18.87
Operating period		
Under 5 years	27	17.0
11–15 years	33	20.8
5–10 years	39	24.5
Over 15 years	60	37.7

(continued)

(continued)

Type of ownership	Frequency	Percent
Labour scale		
Under 10 members	6	3.8
51–100 members	39	24.5
11–50 members	39	24.5
Over 100 members	75	47.2
Revenue		
< 3 billion VND	9	5.7
51–100 billion VND	18	11.3
3–50 billion VND	51	32.1
> 100 billion VND	81	50.9

References

1. Arndt C, Tarp F, Thurlow J (2015) The economic costs of climate change: a multi-sector impact assessment for Vietnam. Sustainability 7(4):1–15
2. International Monetary Fund (2018) For Vietnam, greener growth can reduce climate change risks. https://www.imf.org/en/News/Articles/2018/01/09/NA010918-For-Vietnam-greener-growth-can-reduce-climate-change-risks
3. Tamulis V, Guzavičius A, Žalgirytė L (2012) Factors influencing the use of green logistics: theoretical implications, economics and management. Econ Manage 17(2)
4. Cheng B, Dai H, Wang P, Xie Y, Chen L, Zhao D, Masui T (2016) Impacts of low-carbon power policy on carbon mitigation in Guangdong Province, China. Energy Policy 88:515–527
5. Shen L, Wu Y, Lou Y, Zeng D, Shuai C, Song X (2017) What drives the carbon emission in the Chinese cities—a case of pilot low carbon city of Beijing. J Clean Prod. https://doi.org/10.1016/j.jclepro.2017.10.333
6. Cao K, Xu X, Wu Q, Zhang Q (2017) Optimal production and carbon emission reduction level under cap-and-trade and low carbon subsidy policies. J Clean Prod. https://doi.org/10.1016/j.jclepro.2017.07.251
7. Lyu X, Shi A, Wang X (2020) Research on the impact of carbon emission trading system on low-carbon technology innovation. Carbon Manage. https://doi.org/10.1080/17583004.2020.1721977
8. Mo JL, Agnolucci P, Jiang MR, Fan Y (2016) Impact of Chinese carbon emission trading scheme (ETS) on low carbon energy (LCE) investment. Energy Policy 89:271–283
9. Wang X, Zhu Y, Sun H, Jia F (2017) Production decisions of new and remanufactured products: implications for low carbon emission economy. J Clean Prod. https://doi.org/10.1016/j.jclepro.2017.10.053
10. Lee CT, Hashim H, Ho CS, Van Fan Y, Klemeš JJ (2016) Sustaining the low-carbon emission development in Asia and beyond: Sustainable energy, water, transportation and low-carbon emission technology. J Clean Prod. https://doi.org/10.1016/j.jclepro.2016.11.144
11. Sun L, Cao X, Alharthi M, Zhang J (2020) Carbon emission transfer strategies in supply chain with lag time of emission reduction technologies and low-carbon preference of consumers. J Clean Prod. https://doi.org/10.1016/j.jclepro.2020.121664
12. Wang Q, Zhao D, He L (2016) Contracting emission reduction for supply chains considering market low-carbon preference. J Clean Prod. https://doi.org/10.1016/j.jclepro.2015.11.049

13. Xia L, Qin J, Hai W, Ji F, Yue X (2018) Carbon emission reduction and promotion policies considering social preferences and consumers' low-carbon awareness in the cap-and-trade system. J Clean Prod 195:1105–1124
14. Abbasi M, Nilsson F (2016) Developing environmentally sustainable logistics: exploring themes and challenges from a logistics service providers' perspective. Transp Res Part D: Transp Environ 26:273–283
15. Cantor DE, Morrow PC, McElroy JC, Montabon F (2013) The role of individual and organizational factors in promoting firm environmental practices. Int J Phys Distrib Logist Manag 43:407–427
16. Centobelli P, Cerchione R, Esposito E (2017) Environmental sustainability in the service industry of transportation and logistics service providers: systematic literature review and research directions. Transp Res Part D: Transp Environ 53:454–470
17. Evangelista P (2014) Environmental sustainability practices in the transport and logistics service industry: an exploratory case study investigation. Res Transp Bus Manag 12:63–72
18. Foerstl K, Azadegan A, Leppelt T, Hartmann E (2015) Drivers of supplier sustainability: moving beyond compliance to commitment. J Suppl Chain Manage Inst Suppl Manage 51(1):67–92
19. Goh SH (2019) Barriers to low-carbon warehousing and the link to carbon abatement: a case from emerging Asia. Int J Phys Distrib Logist Manag 49(6):679–704
20. Govindan S, Kalyiyan M, Kannan D, Haq AN (2014) Barriers analysis of green supply chain management implementation in Indian industries using analytic hierarchy process. Int J Prod Econ 147:555–568
21. Lieb KJ, Lieb RC (2010) Environmental sustainability in the third-party logistics (3PL) industry. Int J Phys Distrib Logist Manag 40(7):524–533
22. Lin CY, Ho YH (2011) Determinants of green practice adoption for logistics companies in China. J Bus Ethics 98:67–83
23. Zhu Q, Geng Y (2013) Drivers and barriers of extended supply chain practices for energy saving and emission reduction among Chinese manufacturers. J Clean Prod 40:6–12
24. Gupta P, Anand S, Gupta H (2017) Developing a roadmap to overcome barriers to energy efficiency in buildings using best worst method. Sustain Cities Soc 31:244–259
25. Ibrahim MF, Putri MM, Utama DM (2020) A literature review on reducing carbon emission from supply chain system: drivers, barriers, performance indicators, and pratices. In: IOP Conference Series: Materials Science and Engineering.
26. Jabbour ABLS, Jabbour CJC, Sarkis J, Latan H, Roubaud D, Filho MG, Queiroz M (2020) Fostering low-carbon production and logistics systems: framework and empirical evidence. Int J Prod Res. https://doi.org/10.1080/00207543.2020.1834639
27. Chuai X, Huang X, Wang W, Zhao R, Zhang M, Wu C (2014) Land use, total carbon emissions change and low carbon land management in Coastal Jiangsu, China. J Clean Prod 103:1–10
28. Sun Z, Liu Y, Yu Y (2019) China's carbon emission peak pre-2030: exploring multi-scenario optimal low-carbon behaviors for China's regions. J Clean Prod 231:963–979
29. Naqi A, Jang JG (2019) Recent progress in green cement technology utilizing low-carbon emission fuels and raw materials: a review. Sustainability 11(2):537–554
30. Trivellas P, Malindretos G, Reklitis P (2020) Implications of green logistics management on sustainable business and supply chain performance: evidence from a survey in the Greek Agri-Food Sector. Sustainability 12(24). https://doi.org/10.3390/su122410515
31. Qin B, Han SS (2013) Planning parameters and household carbon emission: evidence from high- and low-carbon neighborhoods in Beijing. Habitat Int 37:52–60
32. Payne C, Scarborough P, Cobiac L (2016) Do low-carbon-emission diets lead to higher nutritional quality and positive health outcomes? A systematic review of the literature. Public Health Nutrition 19(14):2654–2661
33. Wang R, Liu W, Xiao L, Liu J, Kao W (2011) Path towards achieving of China's 2020 carbon emission reduction target—a discussion of low-carbon energy policies at province level. Energy Policy 39:2740–2747
34. Wolf C, Seuring S (2010) Environmental impacts as buying criteria for third party logistical services. Int J Phys Distrib Logist Manag 40(1):84–102

35. Hair Jr JF, Black WC, Babin BJ, Anderson RE (2009) Multivariate data analysis. Prentical-Hall International
36. Lah O (2015) The barriers to low-carbon land-transport and policies to overcome them. Eur Transp Res Rev. https://doi.org/10.1007/s12544-014-0151-3
37. Bhandari RR (2014) Impact of technology on logistics and supply chain management. J Bus Manage 19–24
38. Report Linker (2020) Vietnam freight and logistics market—growth, trends, and forecast (2019–2024). https://www.globenewswire.com/news-release/2020/07/28/2068859/0/en/Vietnam-Freight-and-Logistics-Market-Growth-Trends-and-Forecast-2019-2024.html
39. Phu D, Huong M (2019) 'Độc quyền' cấp nhiên liệu CNG, nguy cơ xe buýt rối loạn. 'Exclusive' of CNG fuel supply, risk of bus disorder. Tuoitre Online.2019. Vietnamese
40. Ministry of Transport (2015) Circular 48/2015/TT-BGTVT. Regulations on registration of inland waterway vessels
41. Ministry of Transport (2020) Circular 26/2020/TT-BGTVT. Amendments and supplements some articles of the occurrences provisioning the regular report regulation in the registration area
42. Ministry of Transport (2016) Decision 1456/QĐ-BGTVT. The issuance of action plan to respond to climate change and green growth of the ministry of transportation for 2016–2020
43. Ministry of Transport (2016) Decision 4206/QĐ-BGTVT. Decision to promulgate the Plan to reduce CO2 emissions in Vietnam's civil aviation activities for the period 2016–2020
44. VILAS (2020) Thực trạng chi phí Logistics Việt Nam năm 2020. Actual situation of logistics costs in Vietnam in 2020. Vietnam Logistics & Aviation School, Vietnamese
45. Anh L (2019) Cước vận tải biển sắp tăng mạnh vì nhiên liệu sạch. Ocean freight rates are about to increase sharply because of clean fuel. The Saigon Times Vietnam, Vietnamese
46. Arndt C, Tarp F, Thurlow J (2015) The economic costs of climate change: a multi-sector impact assessment for Vietnam. Sustainability 7(4):4131–4145
47. Qian C, Wang S, Liu X, Zhang X (2019) Low-carbon initiatives of logistics service providers: the perspective of supply chain integration. Sustainability 11(12):1–13

Development of Innovative Rail Wagons for Transport of Municipal Waste "EKO-VAKO"

Dražen Vidović, Mladen Nikšić, and Monika Šušak

Abstract The paper deals with the issue of transporting municipal waste from the place of origin to the recycling centres by using rail transport. Since waste management is one of the crucial economic branches of the future, both from the aspect of circular economy and the aspect of environmental protection, all the components, including transport, represent increasing challenges in the everyday requirements for an acceptable transport solution in the freight transport. Relevant scientific and professional papers in the field of studying transport capacities in waste management by using rail transport have been analysed. Apart from the problem of transport capacities, also the papers in the field of environmental protection have been processed, which are directly included in solving the issue of environmental protection from the transport pollution. These papers and their conclusions and methods will be applied in the selection and definition of the necessary criteria that will help in later research in defining and designing of a completely new, innovative product, freight rail wagon which will multiply improve the efficiency of the municipal waste transport. The final part of the paper presents an overview of the literature in the field of rail industry, freight wagons for the transport of various types of goods and the ecological aspects of rail transport as a possible solution that may be used in the transport of municipal waste by rail in an efficient, cost-effective and ecologically acceptable manner.

Keywords Transport of municipal waste · Environmental protection · Railway · Innovation · Freight wagon

D. Vidović
Centar za promet i logistiku, Borongajska cesta 83a, Zagreb, Croatia
e-mail: drazen.vidovic@cpl.hr

M. Nikšić (✉) · M. Šušak
Faculty of Transport and Traffic Sciences, University of Zagreb, Vukelićeva 4, Zagreb, Croatia
e-mail: mladen.niksic@fpz.unizg.hr

M. Petrović et al. (eds.), *The Science and Development of Transport—ZIRP 2021*,
https://doi.org/10.1007/978-3-030-97528-9_10

1 Introduction

The transport of municipal waste relies not only in the Republic of Croatia but also in the entire EU, dominantly on road transport, i.e. use of specialized road transport means in the waste transport.

Of all transport modes, road transport is considered the biggest environmental polluter, and it also causes a number of related negative external costs. This refers, concretely, to the adverse impact on the road traffic safety, the necessity of introducing special traffic regulations (especially during tourist seasons) and to the obstruction of appropriate process of corridor traffic management, which is one of the most complex organizational activities in the advanced inter-urban transport.

Besides selecting the technology of waste disposal which would be applied in the waste management centres themselves, one of the crucial challenges is the model and method of economical, efficient, effective and harmless waste transport from the place of collection to the place of its disposal and processing. Accordingly, the main requirements that have been set for the waste transport system include:

- minimal price of waste transport;
- maximal quantity of waste that can be transported on one route;
- efficiency and simplicity of waste manipulation at the points of loading and the points of unloading;
- minimal impact on regular traffic flows (e.g. on the percentage of the congestion of road routes);
- minimal adverse ecological footprint.

In accordance with [1–3], the introduction of an integrated waste management system has been planned. More precisely, the building of a waste management centre (further in the text WMC) has been planned on the county/regional level, as the most important infrastructural facility of waste management. Parallel with this, all the existing landfills in the Republic of Croatia will be remediated and closed. Also, in the process of accession to the EU in accordance with [4, 5], which has been integrated today in [6], the Republic of Croatia has committed herself to introduce an integrated waste management system, in which the WMC represents the critical infrastructural points. Since the building of the WMC is in its initial phase, i.e. the entire infrastructural system is in the phase of preparation and construction, it is the right time to address the challenges expected in the phase of its full implementation in the field. This paper will present the previous studies in the field of waste management, especially the municipal waste, and the necessity of adequate disposal and recycling of the municipal waste. Section 2 analyses the available literature of the comparison of railway transport of goods with other modes of transport. Section 3 of the paper will analyse the available literature that refers to the very types, i.e. series of freight waggons. Section 4 gives an overview of the studies related to innovative freight waggon. The analysis includes the available databases of the innovative solutions on the market of municipal waste transport by railway infrastructure and the potential in future protection of intellectual property and product commercialization.

2 Waste Management

All products and services affect the environment, from the extraction of raw materials for their production, the production process, distribution, use and disposal. These include energy and resource consumption, soil, air and water pollution, and greenhouse gas emissions. Considering the lifecycle involves looking at all stages of the product life to find where it can be improved to reduce its environmental impact and the use of resources. The key goal is to avoid actions that transfer negative impacts from one phase to another. Regardless of whether reused, recycled, incinerated or disposed of at the designated areas, household and industrial waste management comes at a financial and environmental cost. First, waste must be collected, sorted and transported before being treated which can be expensive and may result in greenhouse gas emissions and air, soil and water pollution. A modernized approach to waste management means deviation from considering the waste as unwanted burden and looking at it as a commendable resource [5]. The targets for the EU member states are to recycle 50% of municipal waste and 70% of construction waste by the year 2020. The EU has set a target of becoming a smart, sustainable and inclusive economy by 2020 by implementing a set of policies and measures aimed at a low-carbon economy and efficient use of resources. According to [7], the aim is to insure safe transport of all types of waste, including also hazardous waste. Resource efficiency is one of the basic elements necessary for the creation of a green economy in Europe and further, along with waste recycling efforts, also more sustainable production and consumption patterns.

A total of 220 million tonnes of municipal waste were generated in the EU in 2018. Although this is slightly more than in 2017 (227 million t), it was less than in 2008 (227 million t). From 2005 to 2018, the average amount of waste per capita increased by five per cent in the EU countries, but there are significant differences among the member states. More waste is produced by the richer countries and the countries where tourism is an important industry and a significant source of income. Thus, the amount of waste is growing in Denmark, Germany, Greece, Malta and the Czech Republic, and falling in Bulgaria, Estonia, Hungary, Romania and the Netherlands. In absolute values per capita, Denmark (814 kg), Norway (739 kg), Switzerland (703 kg), Malta (640 kg), Germany (615 kg) and Luxembourg (610 kg) are the first, and Romania (272 kg), Poland (329 kg), the Czech Republic (351 kg), Slovakia (414 kg), Hungary (381 kg), Estonia (405 kg) and Slovakia (414 kg) produce the least waste. In terms of absolute values per capita with 432 kg, Croatia is immediately behind Slovakia [5] (Fig. 1).

The strategic goal defined in [8], the separation of the link between the generation of municipal waste and the economic growth that can be expressed through the household consumption, has not been achieved. In 2018, the GDP in the Republic of Croatia amounted to 12,044 euro/inhabitant, the average net income amounted to 6242 kuna per employed inhabitant, and the amount of produced municipal waste was 432 kg/inhabitant (Fig. 2).

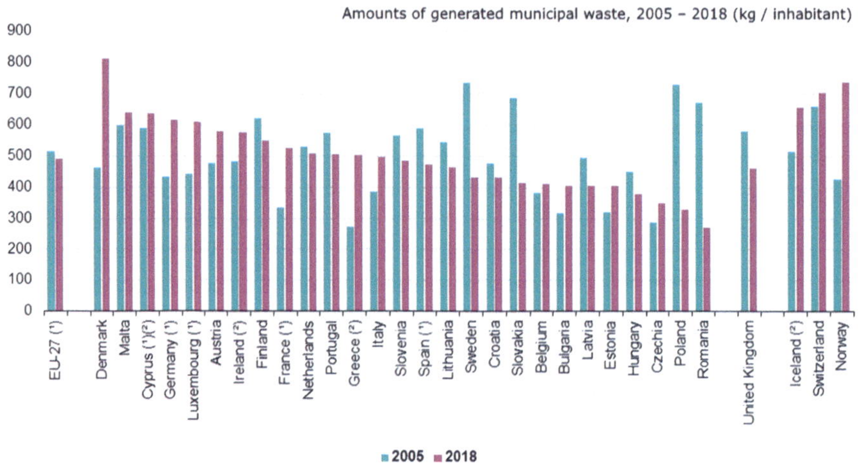

Fig. 1 Annual amounts of generated municipal waste per capita in EU for the period from 2005 to 2018. *Source* Eurostat, 2020, processed by authors

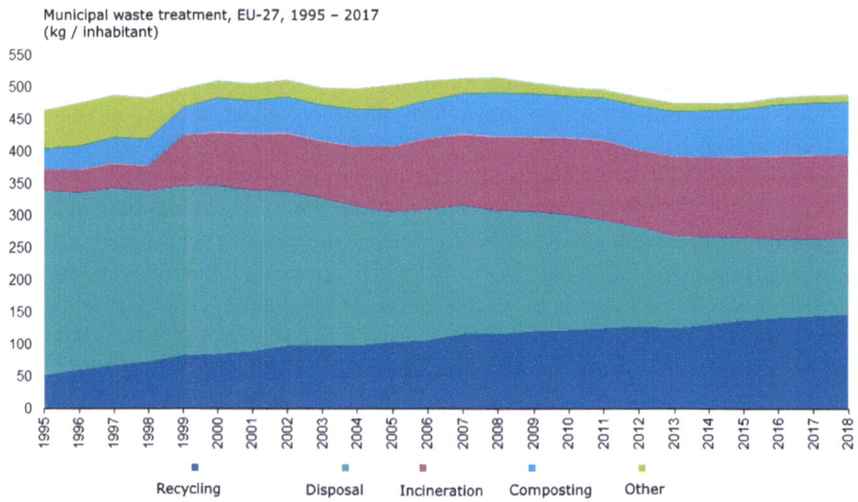

Fig. 2 Share of waste processing/disposal procedures in 2018. *Source* Eurostat, 2020, authors' processing

The citizens' consumption and the overall economic development are growing somewhat faster than the growth of the amounts of municipal waste. Further to [1–3] county waste management centres have been formed whose main focus is the establishment of a new harmonized landfill in combination with the waste pre-treatment procedure, in order to improve the level of waste collection service, to establish separate waste collection systems and to make it possible to close the uncoordinated landfills in the county. Waste management centres have been established with the aim of providing ecologically acceptable, economical and affordable regional bio-degradable and municipal waste management system.

In paper [9], the authors indicate the necessity of identifying the origin and treatment of waste as the basis of circular economy. This study for the city-region of Brussels (Belgium) analyses the municipal waste flow, the intensity of waste generation and the efficiency of waste management. When considering the context of the cities with highly urbanized terrain, it becomes obvious that the closure of material cycles at the city level can only act to a certain extent and is limited to certain flows. Therefore, the visible role of the cities (such as Brussels) in CE is mainly as a supplier of the reusable products, recyclable materials (such as paper, plastic, metal and glass) and nutrients from organic waste. The closure of material cycles at the city level has environmental benefits in relation to the national or global level, needs to be further evaluated based on the comparative LCA studies of cycles at the national or even global level. The authors in paper [10] have developed a model of direct waste collection in big cities, which represents a very challenging problem. The concept of smart waste bins and waste collection in smart cities opens up the possibility of monitoring the amounts of the collected waste in every region and enables timely collection and treatment of the produced waste. This paper introduces the stochastic optimization model in order to improve the procedure of recovering the value of waste collection bearing in mind that the value embedded in the collected waste is uncertain due to various waste mixtures and conditions. The goal is to increase the profit of the recycling process. Apart from maintaining safety (the European railways are among the safest in the world), it is defined in [11] that the goal of EU policy is to maintain a high standard and harmonize the requirements across Europe. In spite of being less polluting for the environment, and safer than road transport according to statistical data, rail transport is failing to compete with it, so the European railways are still not realizing their real potential. The main goal of the European policy is to create a single European transport area that will help Europe to remain competitive by increasing the efficiency of the entire transport sector. In the case of railway transport, the competition lies in road and air transport. The higher the demand for competing products, the lower the prices of rail transport services in order to create higher demand. Another determinant directly affects the achievement of profitability and that is quality. The quality of railway transport services is seen through speed, comfort and reliability. The increase in quality is conditioned by the increase in costs. Technological improvement achieves a higher quality of service. Technological improvements are basically always, in accordance with the profit orientation, aimed with available capital and labour at ensuring the production of a larger amount of outputs at lower costs.

3 Railway Technology—Freight Wagons

In response to intense road competition and changes in the types of freight, there is an increasing desire to produce wagons in such a way as to increase their service life, in accordance with the policy of "greater value for money". This is advantageous regarding the operative and unit costs during the vehicle service life, but it has a negative effect of slow development and innovations in the rail transport sector. Comparing railway and road vehicles, the service life of a rail wagon is much longer than that of a road vehicle (truck). This shorter service life of road vehicles enables faster introduction of new technologies and innovations. The challenge lies in how to establish a balance, so that the manufacturers and the users of the future railway vehicle can achieve "value for money" in investing into a new rolling stock but at the same time adapt to the future technical and operational developments and innovations. The authors in paper [12] analyse the current state of the rolling stock of the Croatian Railways. They propose a structure and dynamics of purchasing new vehicles for the needs of passenger and freight transport. The basic projects that need to be implemented for the purpose of optimizing the rolling stock include programs of modernization of the existing and procurement of new vehicles. These programs have to be mutually harmonized since their realization, regarding the complexity and financing method, will take place in part in parallel. The innovations and especially the ecological innovations as the basic projects that need to be realized for the purpose of efficient use of resources and circular economy by investing into advanced technologies and the development of innovative products of railway industry will enable more efficient, environmentally friendlier and economically more cost-efficient transport and shifting the freight from road to rail. In paper [13], the author presents a new, innovative wagon with a low rotating loading platform for the transport of railway trucks developed at the Military Technological University in Warsaw. The essence of such reloading is the placing of the semi-trailer on a special rotating platform. The construction can be used for the transport of various types of vehicles such as tractors, trucks, trailers, semi-trailers and freight containers. The wagon facilitates fast loading and unloading without the need for any platform or terminal. It was built for intermodal transport system based on innovative concept of a railway wagon. This type of rail wagon will allow the transportation companies to save time and money for road transport. The advantages of such a construction are the reduction of negative impact on the environment as well as increase of safety on the roads by reducing the number of vehicles on the roads. In paper [14], the authors present the results of an environmental assessment of a self-propelled wagon for bulk freight (SPBC), innovative, new product whose aim is to reduce the impact of the railway infrastructure maintenance on the environment. The novelty of the research includes the fact that this is the first study to assess the lifecycle impact of bulk wagons on the environment, as well as the fact that the authors proposed the use of a new functional unit in the category of freight rail vehicles. The paper uses the method of the product life cycle assessment (LCA), and the environmental impact is presented through the categories of impact based on the CML 2001 method through three main lifecycle modules of the self-propelled bulk wagon: upstream, central and downstream. The self-propelled bulk freight wagon has the greatest impact on the environment through its lifecycle in the phase of use and maintenance, in all five impact categories and mainly due to the use of diesel fuel. SPBC uses much less fuel than a conventional diesel locomotive and is therefore more environmentally friendly.

4 New Innovative Wagon "EKO-VAKO"

The transport research and innovation policy should increasingly support in a consistent way the development and introduction of key technologies necessary to develop the European transport system into a modern and efficient user-friendly system. To make it more effective, it is necessary to complement the technological research by a systemic approach, which will take into account the infrastructural and legal requirements, coordination of several stakeholders and large pilot projects to foster acceptance on the market. In order to analyse the success of the ecological innovations in the EU-27 member states, the eco-innovation index is defined, and it is the indicator of the success of ecological innovations in the EU-27 member states. It is developed by applying 16 indicators grouped in five areas: investments, activities, results, socio-economic results and resource efficiency results. The results of each country are compared with the EU-27 average of 100. Insight into the development of eco-innovations promotes the holistic approach to linking economic, environmental and social trends towards sustainable transition of society [15]. In 2018, the largest eco-innovation index was in Luxembourg (138) and Germany (137) and Sweden (132). Eleven states had a value of around 100 of the EU-27 average, and they were labelled as "average bearers of eco-innovations" with the results ranging from 112 (Italy and France) to 88 (Croatia). Observing the results of Croatia in 2018 compared to 2017, a progress was observed in four areas, especially in the socio-economic results (2017—95; 2018—140). Only the results of resource efficiency are in a slight decrease, from 97 in 2017 to 92 in 2018, which indicates the need to improve the productivity of material and energy and reduce the intensity of greenhouse gas emissions. Only water productivity has been significantly improved, from 133 in 2017 to 251 in 2018. The authors in paper [16] analyse the state of the freight rolling stock in Croatia and in the surrounding countries and define the transport and technological demand for new, innovative special wagons for the transport of bulk cargo. It is precisely the special wagons, for special purposes, of innovative technologies, that are required by the market in order to make the railway transport service more efficient, more economical and more acceptable by the market. Based on the conducted research [17], the industrial applicability of the present invention is unquestionable. The search was performed using Cooperative Patent Classification (CPC), a classification coupled with key words on the Espacenet patent database containing about 80 million patents from some eighty countries—downloadable in "full facsimile" form. The search was also performed on the documents that an expert in the field can find in the literature on Google Scholar; Web of Science, ScienceDirect, CROSBI and other similar services. The analysis has been made regarding the novelty and inventive step of the invention and according to the state-of-the-art documents of the technology condition and standards. The review of applications [18, 19] was used to analyse the existing solutions of municipal waste transport protected by a patent application. The majority of the documents is related to improvements related to standard road vehicles intended for waste collection and not to the wagons for its transport, mainly because the rail traffic movement has been clearly defined by the railway network. Development of innovative rail wagon for transport of municipal waste is called EKO-VAKO, which is an abbreviation of the word in the Croatian language (EKOloški VAgon za prijevoz Komunalnog Otpada, eng. ecological wagon for the transport of municipal waste). Thus, the patentability should be sought in the synergy of the "transhipment" methods and "cargo compression" in the wagon. In this sense, the wagon development should be set in the following way:

1. two or several cases on a wagon which can be but do not have to be tilted for unloading at the final destination, if a reverse auger is used for unloading; the technology is already being applied in the existing patented wagon;
2. that the cases are mounted on a bogie, for example:
   which would enable loading into the case from all sides since the case would be turned towards the truck, and there would be no need for loading ramps;
3. that an opening or a mechanism for direct loading and compression of material inside the case is built into the head of the case;
4. that the wagon is self-propelled, as an added value.

Thus, similarly designed wagon, with relatively high probability, has the basis for patent protection in the sense that it combines the rotating case with a single inlet, which at the same time chops and compresses the waste and allows its emptying (Fig. 3).

Fig. 3 Future design of innovative wagon for the transport of municipal waste. *Source* processed by authors

5 Concluding Considerations

By analysing the available literature, it has been found that there is a problem with municipal waste transport from the place of origin to the recycling centres by using rail transport. The profitability of rail transport in the world has been conditioned by constant investments. The railway is becoming increasingly recognizable as the appropriate mass transport means. Railways can be used for the transportation in a short time of large amounts of cargo to necessary locations. The awareness of the needs for the development of railway transport is yet to come to the fore in Europe, and it can be seen that the European Union commissions are trying to put together best possible programs and strategies to create a single transportation network throughout Europe. The investments so far indicate inconsistencies and neglect of this sector. The environmental friendliness, speed, networking and transportation of large amounts of goods are the goals that are planned to be achieved by investments into railway. The problem occurs in the adjustment of the system. Long-term neglect of the railway sector and the politically eligible businesses have resulted in the destruction of almost all potentials for the development of railway transport. Due to the high degree of regulation and standardization of the specifications, the rail vehicles have to meet for the new technological solutions and innovations are scarce indeed, which is why rail transport is considered a very conservative market segment. The review of relevant papers in the mentioned area aims to compare the technical requirements of the system of the existing solutions—road transport of municipal waste by trucks, and above all to determine whether the development is the key to achieving profitability. Road transport is characterized by high cost (on average 5 × higher than the cost of rail transport), the amount of freight that can be transported by specialized trucks is maximally 15t, whereas by rail, it is up to 60t, waste collection mode requires a series of accompanying time and manual operations (loading, unloading) that generate additional cost. The recognition of the problem regarding the lack of capacity of municipal waste transport to the recycling yards by rail is becoming an increasingly important problem because of the constant increase of the road passenger and freight transport volume and because of the common transport policy which encourages the transition of passengers and goods to more economical and ecologically more friendly transport modes, in this case, the railway. In order for the transport of municipal waste as part of the entire process of waste management to be at a satisfactory level, different authors analyse various organizational solutions that refer to the type and organization of transport. All these analyses provide only partial solutions that are based on avoiding investment solutions since they require significant financial means and significant time periods for their realization from planning, preparation of project documentation to the actual implementation and supervision of the project.

Increased investments allow for more developed technological aspects that then require high-level products as well. All this together generates a supply of services that meet the demand and the need of the users and automatically result in higher profits. The investments into mobile capacities are necessary, and the absence of investments in the recent years has reduced competitiveness. Given the significant financial means necessary for the investment, the results themselves will not be visible immediately but rather after a longer period of time. In addition to processing of the papers that deal with the issue of organizing the municipal waste transport, the paper provides examples of the review of papers related to the comparison of the transport modes, in order to find the most acceptable solution in municipal waste transport in terms of transport efficiency, environmental impact, safety, economy and efficiency of the entire waste management process. The papers have also been processed that give an overview of the freight rolling stock with a recommendation of which series and for which types of transport they will be needed in future. The goal after reviewing the relevant papers in the mentioned areas is to develop a completely new, innovative wagon for the transport of municipal waste. This refers primarily to the transport of municipal waste over greater distances where rail transport is the most cost-effective mode of transport. This means that there is currently on the market no similar type, series of freight wagon that would be intended for the transport of municipal waste. The capacity of such a method of transporting municipal waste is multiply higher than truck transport, whereas on the other hand, the emission into the environment when transporting the same amount of waste is many times lower. Additional research is required so that the technical characteristics of the freight wagon for the transport of municipal waste can be properly defined and so that the characteristics of the developed one would be such that the combined model of municipal waste transport can be properly organized, i.e. to provide accurate input parameters for its proper operation and for the entire system to give satisfactory results. The analysis of the current situation establishes a positive relationship between the real needs, profitability and innovative solutions that drive the economy. This paper represents the basis for future research and the development of innovative products of the railway industry in the field of municipal waste transport that would be based on the eligibility criteria and would be as such applicable in actual operational conditions of the transport process.

References

1. Zakon o otpadu (NN 178/04, 153/05, 111/06, 60/08, 87/09) (Act on Waste)
2. Strategija gospodarenja otpadom Republike Hrvatske (NN 130/05). Waste Management Strategy of the Republic of Croatia
3. Plan gospodarenja otpadom u Republici Hrvatskoj za razdoblje 2015–2022 godine. Plan of Waste Management in the Republic of Croatia for the period from 2015 to 2022
4. Okvirne direktive o otpadu 75/442/EEZ. Chapter 27—Environment. Frame Directives on Waste
5. Good Jobs in the Circular Economy, Vera Weghmann, December 2017
6. Directive 2008/98/EC of the European Parliament and of the Council of 19, November 2008 on waste and repealing certain directives
7. Zakon o održivom gospodarenju otpadom. NN 94/13, 73/17, 14/19, 98/19. Law on Sustainable Waste Management
8. Uredba o pošiljci otpada http://ec.europa.eu/environment/waste/legislation/index.htm Kolovoz 2020. Ordnance of Waste Delivery
9. Strategija održivog razvitka Republike Hrvatske (NN 30/09). Strategy of Sustainable Development of the Republic of Croatia

10. Zeller V, Towa E, Degrez M, Achten WMJ, Urban waste flows and their potential for a circular economy model at city-region level. Institute for Environmental Management and Land-Use Planning, Université Libre de Bruxelles (ULB)
11. Shah PJ, Anagnostopoulos T, Zaslavsky A, Shah SB (2018) A stochastic optimization framework for planning of waste collection and value recovery operations in smart and sustainable cities, 11 May 2018
12. Nikšić M, Mlinarić TJ, Brkić M (2012) Utvrđivanje Prometne i Tehnološke Potražnje Za Željezničkim Vozilima u Republici Hrvatskoj
13. Tadeusz N (2016) Innovative project of prototype railway wagon and intermodal transport system. Military University of Technology, Gen. Sylvestra Kaliskiego St. 2, 00-908 Warsaw, Poland, 18–21 Apr 2016
14. Hegedić M, Štefanić N, Nikšić M (2018) Assessing the environmental impact of the self-propelled bulk carriage through LCA
15. Europska komisija, Opservatorij za eko-inovacije, Uspješnost Hrvatske u eko- inoviranju u odnosu na prosjek EU-28 u 2018. Godini. https://www.eea.europa.eu/themes/sustainability-transitions, Kolovoz 2020
16. Utvrđivanje prometne i tehnološke potražnje za specijalnim vagonima za prijevoz rasutog tereta, Fakultet prometnih znanosti, Zagreb (2011)
17. Europski patentni ured (EPO) (2020) https://www.epo.org/lawpractice/legaltexts/html/guidelines/e/g_vii_7.htm. Kolovoz 2020
18. Međunarodna prijava patenta EP0548295: WO93/1112 (1993) Refuse collection vehicle with a removable collection container, process for removing the refuse, ejection system for emptying the container
19. Međunarodna prijava patenta DE3042046 (1983) Transferring collected refuse to rotary container—involves hopper with pusher and fixed blades to cut up refuse via blades in container entry

Assessment of the Impact of Technological Development and Organizational Complexity in Air Transport

Valentin Marian Iordache, Sorin Eugen Zaharia, and Casandra Venera Pietreanu

Abstract The complexity of the organization–technology–human factor relationship can give rise to a chain of events that are difficult to manage during operational phases. In this regard, the research addresses the elements that can affect the organization–technology relationship whose impact can be determined (from a safety perspective) by the influence of the human factor. After performing a critical review of the literature regarding organizational safety and the current stage of technological development, the paper approaches the subject of organizational behavior managed toward improved adaptability and safety risk minimization. Since probabilistic risk assessment is part of defining the level of safety risks, the authors propose a mathematical model meant to evaluate the probability of operational risk based on the reliability of aeronautical systems. Applying a logical argumentation on the need and importance of both theoretical and practical training, the research provides an organizational management perspective. Through the values obtained following the reasoning and mathematical formulation proposed in this paper, conclusions can be drawn to reduce operational risk through training, thus providing an opportunity to address the issues for air operators in order to maintain a safe and high operational level.

Keywords Aviation · Efficiency · Organizational culture · Risk assessment · Safety · Technological development

1 Introduction

Systems complexity is not a new subject. It was well documented in the last decades due to aeronautical systems' continuous development, and different risk models were created to cover the gaps that older/traditional models were not able to identify thoroughly during analysis. A model that studies the interactions between systems is the system-theoretic model processes (STAMP method), and based on this approach,

V. M. Iordache · S. E. Zaharia · C. V. Pietreanu (✉)
Faculty of Aerospace Engineering, University Politehnica of Bucharest, Polizu Street 1-7, Sector 1, 011061 Bucharest, Romania

M. Petrović et al. (eds.), *The Science and Development of Transport—ZIRP 2021*,
https://doi.org/10.1007/978-3-030-97528-9_11

multiple risks can be identified during systems evaluation [1, 2]. The systems complexity problematic in aviation is depicted also by ICAO's DOC 9859 (Safety Management Manual) [3], which briefly describes in Chap. 2, the evolution of safety, presenting which factors (technical, human or organizational) have had impact on the safety state throughout the history of air transport and how they evolved.

However, it is a mistake not to take into account that these factors are also influencing each other. According to STAMP model, control changes as systems are developed; therefore, the interactions are different and may create new hazards [4]. Wang (2017) presented that in aviation, different types of failures have an acceptable rate [5], thus in order to maintain a high safety and operational efficiency state at all time, and understand the needs of control processes and how to develop improved organizational culture, analyzing the occurrence risk probability is imperative.

The benefits of modern aviation technology are conspicuous and represent the main pillar that continues to lead to the development of technical or organizational aeronautical systems. But the debate over who should have (the final) control—the man or the machine—remains an interesting and much debated topic, as will be presented in the analysis of the two competitors: Boeing and Airbus. This topic has different features and implications depending on the operational aeronautical environment, but the issue of replacing human decision with automation must be operational delimited [6].

The safety level might sometimes be a result of organizational decisions, which directly affects operational activity. Within the organization, safety risks may depend on several internal structures and without good communication, involvement and sufficient investment, modern technologies can produce catastrophes [7]. Whether we are talking about airports, aircraft or all the other processes required for flight, technology has brought a high level of safety, efficiency and reliability among both operators and users of air transport, but the real issue is to be properly implemented and managed throughout operational phases and during the product's life cycle [8, 9].

After studying the modern theories that underlie the understanding of systems operations in air transport and the specific elements of aviation safety management system (SMS), the paper aims to present a new approach for assessing the safety risk probability of interacting systems, based on their individual system reliability (i.e., organizational, technological and human). Through the process of obtaining a numerical reference for risk probability, aspects related to the complexity and importance of control elements were clarified.

1.1 Methodology

The research method partially comprises an analysis of secondary data gathered from specialized organizations: International Civil Aviation Organization (ICAO) safety indicators, ICAO Annexes 13 (Aircraft Accident and Incident Investigation) and 19 (Safety Management) and Safety Management Manual (SMM-ICAO Doc 9859), Federal Aviation Administration (FAA) data and investigation/prevention policies, airworthiness directives (AD) and service bulletins (SB), the reports of the authorities responsible for air transportation accident investigation (the National Transportation Safety Board (NTSB)) and interviews of aviation experts. Further, a review of various scientific articles and surveys on the current stage of technological development was performed, and reports on the technological evolution of the Boeing 737 aircraft over time were studied. In this regard, the authors' own critical apparatus was used for interpreting how new construction elements might alter the patterns of commercial air operation safety and efficiency.

The study of Boeing 737 MAX 8 accidents was meant to identify how information gap can affect training, process planning and decision-making, and emphasized the manner in which systems interact and alter safety levels, thus outlining that complexity paradigm becomes the paradigm of organizational theories. This analysis represented the benchmark for the calculation model of the operational risk probability.

2 Transformation of the Organizational Perspective in Air Transport Due to Technological Development

"Complexity" has become one of the most important words in modern organizational research [10] after different systems have greatly developed within organizational theory. As systems are consistently developing, with the increase in technological/software resources, new risks may arise and operators' knowledge becomes limited. In highly competitive environments, organizations experience more nonlinear and dynamic behaviors; Yang et al. [11] showing that organizations and their environment are in correlative dependence and interaction, while Scott (2003) [12] defining the impact of the environment on the organizational structure as the "interdependent cycle" model.

In a complex environment, strategic goals might help the organizations gain advantages over competitors. Good strategies define the rivalry between Boeing and Airbus; both companies having government contracts, political trust and unparalleled historical depth, surviving in the field of commercial air transport unlike other manufacturers, such as CONVAIR or Lockheed Martin. Using institutional opportunities to improve their own performance, Boeing and Airbus increased productivity and developed new technologies that produced changes in their environmental and organizational practices, forcing them to evolve continuously, while having different philosophies. Airbus aircraft are built so that the computer is in control, not the pilot, while Boeing's philosophy in manufacturing is defined emphasizing the role of the pilot [13]; philosophy embedded in the design of their fly-by-wire system that allows the pilot's full "authority" over the aircraft, regardless of the fact that following the operation of controls, the aircraft is diverted from its flight tire.

The latter approach meant, however, that inadequate pilot training, insufficient experience and regulations and also management limitations in the context of implemented new technologies (e.g., the MCAS—even though destined to improve safety) have proven to be the reason for Boeing's 737 MAX 8 aircraft worldwide grounding. So, technology plays an important part in aviation, because it provides opportunities to enhance the human–machine interface.

The situation with Boeing 737 MAX 8 is different from the usual control systems philosophies we have encountered throughout the years in aviation; flight computers may limit the control inputs of pilots to maintain the flight envelope, but in the situation of the two accidents with MAX 8, the computers had priorities over the flight controls, when the controls where in pilot control mode (manual). The computers intervened in certain situation, not very thoroughly explained to the pilots, and the inputs where from sensors that could have sent wrong data about flight attitude.

This shows that acquiring and maintaining a level of technical knowledge is related to the development of various skills (physical, mental and interpersonal), but also to procedural knowledge; Pietreanu and Iordache (2020) indicating that technological complexity in the aeronautical system should be approached as a network of knowledge [14].

Nevertheless, it is difficult to define levels of application of technical knowledge held during operational processes, tasks are accomplished through skills being directed by a set of standard procedures. In circumstances for which the skills and procedures learned are not sufficient (defined by Rasmussen as Skill-Rule-Knowledge) [15], additional technical knowledge correctly understood might be required. Defining a correct safety performance management should consider the following aspects (Fig. 1):

The air investigation report of the two Boeing 737 MAX accidents (Lion Air Flight 610 and Ethiopian Airlines Flight 302) highlighted seven major safety issues [16, 17], bringing to the fore the crew's inability to manage the situation due to task complexity, confusing and stressful environment (see Table 1) caused by multiple conflicting warnings, doubled by unintentional ignorance of repetitive commands that came from the MCAS (Table 2).

In the case of the MCAS, the level of available information should have balanced the two principles of "need-to-know" and "nice-to-know". However, the risk assessment of the new system functionality (see Table 2) shows that ultimately, the organizational environment is the one that encompasses all forces that may impact performance or operations.

After the first accident, Boeing's delayed reaction was considered to be attributed to the unexpected impact, since the producer previously recorded only few events; safety always being a strong feature of the manufacturer's products, in the case of the 737 aircraft including [18]. This shows one more time that organizational behaviors should be managed toward improved adaptability through two levels: adaptive components of the basic systems and the structure of the systems that fold on these components.

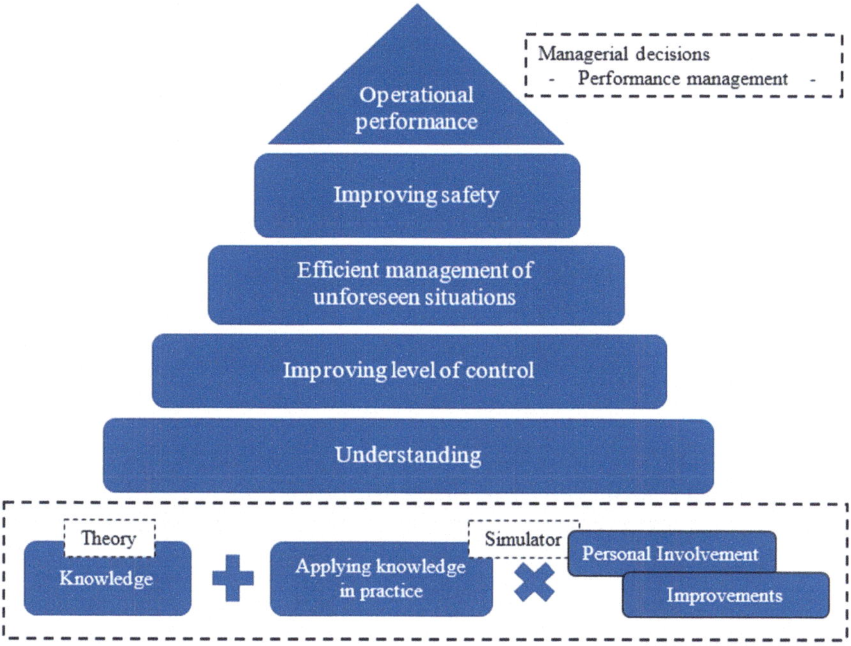

Fig. 1 Organizational performance defined by safety knowledge

Table 1 Safety culture approach on Boeing 737 MAX accident

Safety/organizational culture identified issues	Characteristics
External pressure	Threat represented by the launch of Airbus A320neo (a rival of B737 with 15% increased fuel efficiency)
Organizational/Management issues	Ignorance of pilot's previous complaints/reports regarding the MCAS
Pilot's competence associated issues	Failure to provide adequate simulator training on MCAS Flight manuals lacked detailed documentation
Procedural issues	FAA accreditation approval of MAX under assumption that the aircraft was a derivative of the original 737
Regulatory issues	Incomplete/non-transparent information about MCAS functionally to regulator

Source Adapted after [18]

Table 2 A risk assessment of the MCAS functionality

Risk analysis	Characteristics
Hazard	Lack of correct/complete information of MCAS functionality
Consequences	Automatically reactivation of MCAS after 10 s disconnection Multiple conflicting warnings: overriding the pilot's control Lack of awareness/Uncertainty
Proposed control	A third sensor that can help in case of information discrepancy from the two existing sensors Organizational/safety culture improvement
Risk assessment	Intolerable
Mitigation actions	Pilot additional training and briefs/Increase supervision Appropriate regulations
Risk after mitigation	Tolerable (with existing control)

3 Probability Assessment of Operational Risks in Commercial Aviation

The following analysis aims helping air operators understand how risk probability changes due to systems interaction, indicating how it can be reduced through a training system that covers the information spectrum needed in air transport. The model is divided into two stages: The first stage calculates the value of the risk due to the interaction between the systems (environment, technology and human factor), and in the second stage, the importance of theoretical and practical training is demonstrated in the context of the company's organizational risk.

Organizational risk is considered to be the residual value of Jet Airliner Crash Data Evaluation Centre (JACDEC) safety index; value obtained by subtracting the percentage of JACDEC risk index from the maximum possible value, implicitly 100% [19]. For calculations, the human factor was considered as being a system, the environment and maintenance another system that influences the technological one (in this case, the MCAS from Boeing 737 MAX 8 aircraft) (Fig. 2). MCAS was taken as a reference because it was directly affected by maintenance and the environment, allegedly, in the two disasters, and the risk has increased considerably due to elements that would not have affected the aircraft from past generations, under the same conditions.

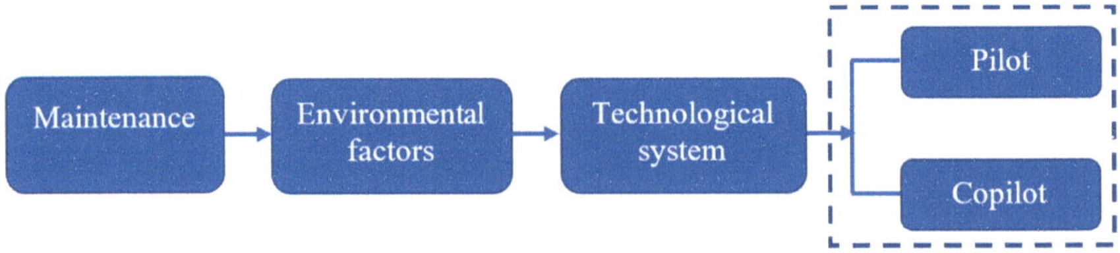

Fig. 2 Systems interaction in the case of the two catastrophes involving Boeing 737 MAX

It is common to use the risk matrix based on risk probability and occurrence, but finding the link between multiple factors that are interacting could provide a better understanding of the need to increase safety and efficiency. The authors managed to obtain a numerical value for risk probability during operational processes by using available data and information from multiple scientific research studies in order to demonstrate the need for developing the systems reliability in aviation.

In terms of systems reliability, the two accidents in which Boeing 737 MAX 8 aircraft were involved will be analyzed. First, the authors presume that all interacting systems have a reliability of 95% (0.95), except for human factor, which is 85% (0.85) for the pilot and 75% (0.75) for the copilot [20]. For the human factor, the analysis considers reliability in parallel, since the tasks are divided between the crew, each pilot having very well established roles, but which can change at any time. For the other systems (i.e., technological and environment), series reliability will be considered, applying the following equations:

$$R_{\mathrm{MAX\,8}} = 0.95 \cdot 0.95 \cdot 0.95 \cdot [1 - (1 - 0.85)(1 - 0.75)] = 0.859 \cdot 0.963 = 0.83 \quad (1)$$

This calculated reliability provides a residual risk probability of 0.17 (17×10^{-2}), value calculated as a result of systems interaction.

If the elements that caused the catastrophes did not influence the other systems (implicitly did not alter safety and efficiency), then we can assume the following values obtained, using the calculation formula for parallel reliability, as follows:

$$R_{\mathrm{MAX\,8}} = [1 - (1 - 0.95)(1 - 0.95)(1 - 0.95)(1 - 0.963)] = 0.999995 \quad (2)$$

In this situation (result 2), the system has a higher reliability, the residual safety risk probability being low: 0.000005 (5×10^{-6}). The series reliability is used to demonstrate how systems are affected by another/each other if they don't have back-up. The parallel reliability is used to demonstrate that even if there is a faulty element that can affect the operation of a system, because of redundancy, that element is acknowledged and removed from the operation cycle so that it cannot influence the safety state. Otherwise, it can cause a chain reaction and affect other good operational systems and have unwanted consequences, as it happened with the MCAS in the Boeing 737 MAX 8 accidents. Thus, in parallel reliability, redundancy is ensured and in series reliability there is a direct influence between systems.

This rearrangement of elements that caused the air disasters demonstrates that an initial design that minimizes the impact of improper maintenance and environmental elements on the MCAS operational system would have provided a much higher reliability to the Boeing 737 MAX 8 aircraft; thus, the associated safety risk could have been low.

Further, the two flights of the MAX 8 aircraft associated with Lion Air (LNI610) and Ethiopian (ETH302) catastrophes, but also one other flight before the accidents (LNI043—flight performed with the same aircraft as flight LNI610, which had

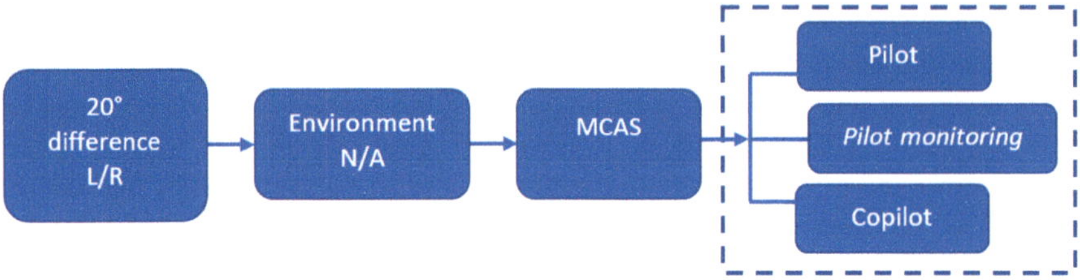

Fig. 3 Systems interaction in the case of flight LNI043

Table 3 Lion Air (2018) and Ethiopian (2019) safety risk ranking

Airline/Country	IATA/ICAO code	No. of serious incidents	Safety ranking position	Risk index (%)
Ethiopian Airlines/Ethiopia	RT/ETH	19	97	85.86
Lion Air/Indonesia	JT/LNI	15	99	55.64

Source [19]

the same issues, but the pilots acted differently), will be considered to analyze the reliability/risk probability values obtained in order to draw final conclusions. The following data (Fig. 3) represent the starting point for analyzing the safety characteristics of the airlines considered in the research, figures placing the two operators at the bottom of the rankings (positions 97 and 99 out of 100). Although for Lion Air the number of safety events was lower, the total number of hull losses since 1989 was higher than Ethiopian's (i.e., 8 vs. 6) and so was the number of non-fatal hull losses (6 for Lion Air and 3 for Ethiopian Airlines) [19] (Table 3).

For MCAS errors during Lion Air flights, we will appreciate a 60% (0.6) reliability, value resulting from maintenance and hitting a bird during take-off with the MCAS sensor, which led to erroneous information received by the system—the sensor has been damaged/not properly calibrated. The value is relative and is considered to be 60% because the system was not reliable, although not completely inoperative; the risk generated by the errors could have been administrated properly.

During the LNI043 flight, the aircraft had the same problems as presented in the 2 Boeing 737 MAX accidents [16, 17], but there were three pilots on board the aircraft, the third having the status of pilot monitoring, who might have helped the crew in solving the situation working together with the pilot and copilot to fix the problem with the aircraft's control systems (Fig. 3).

The human factor (human system) considered in Fig. 3 is arranged according with parallel reliability because it is cooperating to assure an operational process. Even though they were working together, the other systems involved are arranged according with series reliability, since the human system (which is the crew) depends on the impact of the other systems on MCAS. The first element is the influence of 20° difference of angle of attack between left and rights sensors due to poor maintenance;

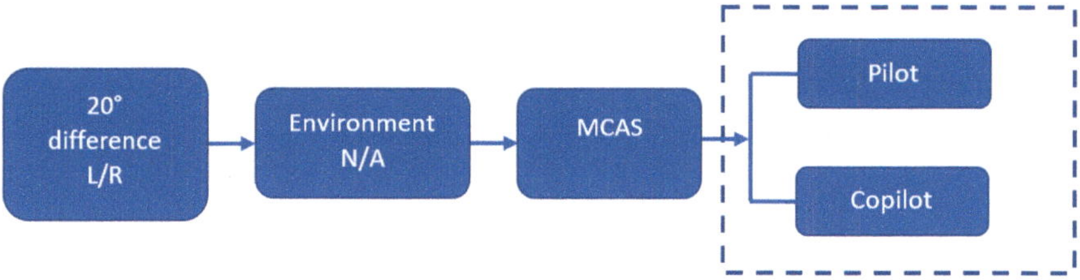

Fig. 4 Systems interaction in the case of LNI610 flight catastrophe

the second is the environment which had no influence whatsoever in this situation on MCAS, and finally the human system who is a key component in air operations process. So the reliability of the interaction between the systems in the case of flight LNI043 will be calculated as follows:

$$\begin{aligned} R_{\text{LNI043}} &= 0.60 \cdot 0.95 \cdot 0.95 \cdot [1 - (1 - 0.85)(1 - 0.75)(1 - 0.8)] \\ &= 0.542 \cdot 0.993 = 0.54 \end{aligned} \quad (3)$$

The reliability is 54% (relation 3), which means that the residual risk is high 0.46 (46×10^{-2}). Next, for describing systems interaction in the case of the LNI610 flight (Fig. 4), the principle of series reliability was accounted.

$$\begin{aligned} R_{\text{LNI610}} &= 0.60 \cdot 0.95 \cdot 0.95 \cdot [1 - (1 - 0.85)(1 - 0.75)] \\ &= 0.542 \cdot 0.963 = 0.52 \end{aligned} \quad (4)$$

The calculated reliability in the case of the Lion Air disaster is lower than the previous case (52%), which means that the residual risk is 0.48 (48×10^{-2}). It can be seen that the difference between the flight LNI043, who had a pilot monitoring on board and LNI610 flight, is a quite small value, but which may be enough to solve an emergency. Also, luck, flair or inspiration should not be excluded in situations such as in the present case, which can be materialized mathematically as a 2×10^{-2} value, in a dynamic and complex system according with the principle applied. If the system was built so that man could take control—so the MCAS did not have priority over commands during flight on manual control [21]—then the systems would have had a different organization scheme, and the impact of the environment and/or maintenance would have been limited.

In this case, the reliability is a combined one; first the series reliability between the MCAS, the operational system (environment) and the maintenance system must be calculated. Afterwards, using the parallel reliability, the reliability of the MCAS is calculated—affected by external conditions—with the human factor having the ability to intervene and correct system errors. In this case, the MCAS is decoupled by the pilot via a switch, and the pilot takes control of the aircraft. Therefore, the calculated reliability is:

$$R_{\mathrm{LNI610}} = 1 - \{[1 - (0.60 \cdot 0.95 \cdot 0.95)] \cdot [1 - (1 - 0.85)(1 - 0.75)]\}$$
$$= 1 - 0.458 \cdot 0.037 = 0.983 \tag{5}$$

Thus, systems interaction reliability (result 5) of the Lion Air flight 610 (if one of the pilots could have intervened to fix the errors of MCAS) is 98.3%, and the residual risk probability is 1.7×10^{-2}.

For Ethiopian Airlines flight ETH302, an accident that occurred four and a half months after the first Boeing 737 MAX 8 catastrophe, the situation was different; there were information about the MCAS (information which, for calculation, has assigned a value of 80%) about how the flight controls are operated and how to operate properly (or to decouple it). However, even with those information, the impact that the system had on the flight situation and the lack of practical training in flight simulator for pilots in complex situations, led to another air disaster [18, 22]. For the flight ETH302, the following form and equation of human factor–system interaction is considered (Fig. 5, Formula 6). The interaction between systems was arranged according to series reliability, and the reliability of the system represented by the human factor is apx. 0.95 (calculus 6).

$$R_{\mathrm{HumanFactor}} = 1 - (1 - 0.85)(1 - 0.65) = 0.9475 \tag{6}$$

Given that the sensors for the MCAS are located in the frontal area of the aircraft, the risk of a bird strike affecting one of the two sensors is high. In order to calculate the probability of hitting the sensor by a bird (as it might have happened in the case of ETH302), the studies of Hedayati and Sadighi [23] were taken into account. In order to obtain a greater accuracy in the calculations, the reported values for hitting the radome and the front panels were considered, as these (together with the engines and wings) are the components that are most affected by birds impact [24].

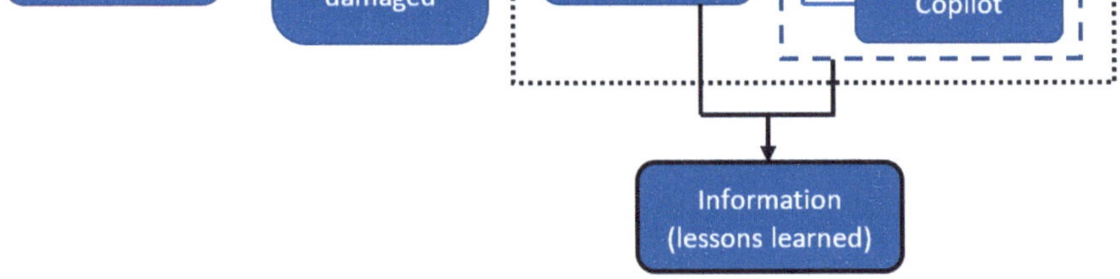

Fig. 5 Systems interaction in the case of ETH302 flight

For the calculations, the value 29.19% will be taken into account, which represents the percentage of minor damages caused by birds being hit/hitting the aircraft. Also, the value 70.81% (100–29.19%) will be used (Formula 7)—this representing the reliability of the environment; and 0.9475 representing the calculated reliability of the human factor (see calculus 6).

$$R_{\text{ETH302}} = 0.95 \cdot 0.7081 \cdot [1 - (1 - 0.95)(1 - 0.80)(1 - 0.9475)] = 0.67 \quad (7)$$

Given the calculated reliability of the system (67%), the residual risk remains 33% (33×10^{-2}). This result is almost the same as in the case of the LNI610 catastrophe, the risk being still present since mitigation actions after the first accident were not appropriate—insufficient and inefficient. After the ETH302 disaster, flights were resumed after FAA certifications on December 29, 2020 (American Airlines) only in the airspace of the USA [25].

In this case, it can be seen that there is no difference from the initial catastrophe, although another component appeared in the system (namely the information system that provided more clarification after the LNI610 catastrophe). But the impact that the environment might have had (so the operating system) by allegedly hitting that bird and which in the calculations has assigned a reliability of 70.81%—the chances of hitting the bird with one of the sensors being quite low, but not impossible—was high enough to adversely affect the flight characteristics of the aircraft, as the pilots were unable to take control of the aircraft due to the way the MCAS was implemented in the 737 MAX 8 aircraft and its poor response.

4 Conclusions

In aviation, safety has become a multidisciplinary issue given the multitude of areas and systems involved that need to work together on an ongoing basis. Therefore, awareness of the differences that arise in operation due to technological development must not be limited to embracing the benefits brought by new systems, but understanding how risk is resized and how a simple element can produce macro effects.

The paper presented the link between the main systems that interact in aviation (organizational, technological and human) and identified the elements that balance the relationships and interactions between them from a safety perspective, in the context of the complexity of aeronautical systems and with the purpose of assessing the probability of risk in operational processes. The high trust in technology and lack of redundancy between systems, which can be demonstrated by using series or parallel reliability, were the main new elements that the two Boeing 737 MAX 8 accidents have proven to represent potential hazards.

The authors outlined that the difficulty of managing risks and the lack of information were the premise for the occurrence of the two air disasters involving Boeing 737 MAX 8. Intricate features of MCAS were not very transparent to pilots, and given the complexity, the information processing capabilities and time constraints were dependent of pilot training. Thus, the authors indicate that technological complexity should be approached as a network of knowledge, vital for maintaining an appropriate level of safety, where necessary information should be sufficiently detailed to cover the concept of "operational-need-to-know".

References

1. Leveson N (2004) A new accident model for engineering safer systems. Aeronaut Astronaut J USA, Massachusetts Institute of Technology
2. Altabbakh H, Alkazimi M, Murray S, Grantham K (2014) STAMP—holistic system safety approach or just another risk model. J Loss Prevention in the Process Industries, 32
3. Thomas J, Leveson N (2018) STPA handbook. system-theoretic process analysis handbook, massachusetts institute of technology, pp 1–188. https://psas.scripts.mit.edu/home/get_file.php?name=STPA_handbook.pdf. Last Accessed 15 Apr 2021
4. ICAO (2013) DOC 9859 Safety Management Manual (SMS). International Civil Aviation Organization, Montréal, Canada
5. Wang P (2017) Airworthiness regulations and safety requirements. civil aircraft electrical power system safety assessment, issues and practices, pp 1–26. https://doi.org/10.1016/B978-0-08-100721-1.00001-7
6. Chialastri A (2012) Automation in aviation, automation, IntechOper, Medicair, pp 79–102, Rome, Italy. doi: https://doi.org/10.5772/49949
7. Aven T (2016) Risk assessment and risk management: Review of recent advances on their foundation. Euro J Oper Res 253(1):1–13. ISSN 0377–2217. https://doi.org/10.1016/j.ejor.2015.12.023
8. FAA (2010) Aviation Safety: Improved Data Quality and Analysis Capabilities Are Needed as FAA Plans a Risk-Based Approach to Safety, GAO report number GAO-10–414, U.S. Government Accountability
9. ICAO (2019) Safety, State of Global Aviation Safety, International Civil Aviation Organization Safety Report
10. Grösser S (2017) Complexity management and system dynamics thinking, dynamics of long-life assets. Springer, Cham, pp 69–92. https://doi.org/10.1007/978-3-319-45438-2_5. Last Accessed 11 Apr 2021
11. Yang CX, Liu HM, Wang XX (2013) Organization theories: from classical to modern. J Appl Sci 13(21):4470–4476. ISSN 1812-5654, http://docsdrive.com/pdfs/ansinet/jas/2013/4470-4476.pdf. Last Accessed 25 May 2021
12. Scott WR (2003) Organizations and organizing: rational, natural and open systems, 5th edn. Pearson Prentice Hall, Columbus, Ohio
13. Spitzer C, Ferrell U, Ferrell T, Becker SG (2015) Digital avionics handbook. CRC Press, Taylor & Francis Group, London
14. Pietreanu CV, Iordache VM (2020) Limitations of organizational and risk management performance in air transport due to technology and the dynamics and unpredictability of the global environment, In: Proceedings of the 7th international management conference-review of management and economic engineering "RMEE", 17–19 September 2020, Cluj, Romania. ISSN 2247 – 8639, ISSN-L = 2247 – 8639; 2020, pp 181–186
15. Hobbs A, Williamson A (2002) Skills, rules and knowledge in aircraft maintenance: errors in context. Ergonomics 45(4):290–308. https://doi.org/10.1080/00140130110116100
16. FAA (2020) Summary of the FAA's review of the Boeing 737 MAX, return to service of the Boeing 737 MAX aircraft, federal aviation administration

17. NTSB (2019) Safety recommendation report, assumptions used in the safety assessment process and the effects of multiple alerts and indications on pilot performance, National Transportation Safety Board, Washington, DC 20594
18. Naor M, Adler N, Gavriel GD, Dumanis A (2020) Psychological safety in aviation new product development teams: case study of 737 MAX airplane. Sustainability 12:8994. https://doi.org/10.3390/su12218994
19. JACDEC (2019) Airline safety ranking 2019, Airline Ranking
20. Earnhardt PM, Newcomer JM, Watkins DV, Marion JW (2014) An inquiry into the aviation management education paradigm shift. Int J Aviat Aeronaut Aerosp 1(4). (Embry-Riddle Aeronautical University, SUA)
21. Boeing (2021) 737 MAX-MCAS, 737 MAX maneuvering characteristics augmentation system. http://www.b737.org.uk/mcas.htm. Last Accessed 21 May 2021
22. EASA (2021) Boeing 737 MAX return to service report, overview of the technical investigation activities performed by EASA
23. Hedayati R, Sadighi M (2015) Bird Strike: an experimental, theoretical and numerical investigation. Woodhead Publishing in mechanical Engineering. https://www.researchgate.net/publication/302181762_Bird_Strike_An_Experimental_Theoretical_and_Numerical_Investigation. Last Accessed 8 apr 2021
24. Metz I, Ellerbroek J, Mühlhausen T, Kügler D, Hoekstra J (2020) The bird strike challenge, aerospace, MDPI
25. American Airlines (2020) 737 MAX return to service, AA report

Strategic Settings for the Development of Franjo Tuđman Airport Cargo City as a Regional Centre

Jure Šarić, Andrija Vidović, Igor Štimac, and Tomislav Mihetec

Abstract In an age of globalized economy and very dynamic competition for economic activities worldwide, airports are increasingly presented as strategically the most important transport infrastructure, expanding economic activity from capitals to the urban periphery and airport surroundings. Existing airports are no longer just transport infrastructure, but are becoming multifunctional complexes, whose impact on the surrounding region could be significant in various ways, primarily in business and commercial terms. Access to the airport is an important criterion for selecting the airport, and, thus, for its competitiveness. The development of the infrastructure for receiving and dispatching Cargo at the Franjo Tuđman Airport in Zagreb, Croatia, refers not only to airport facilities and spaces, but also to the larger logistics area surrounding the airport, including access to multimodal transport infrastructure. The role of services in modern economies has significantly increased, in terms of their production, employment and importance, as inputs for other sectors, which requires additional investment in modern infrastructure for receiving and shipping cargo with aim to strength competitiveness and position in the cargo market. This paper analyses the existing infrastructure's capacities and estimates future needs for air cargo transportation in the catchment area of the mentioned airport. Furthermore, the paper presents a conceptual and methodological framework for the development of airport infrastructure and its impact on regional economic growth. In order to better present the results of the work, various methods were used, such as the method of analysis and synthesis, comparisons, statistics and descriptions. The research aims to determine

J. Šarić (✉)
Croatian Civil Aviation Agency, Ulica grada Vukovara 284, 10000 Zagreb, Croatia
e-mail: jure.saric@ccaa.hr

A. Vidović · I. Štimac · T. Mihetec
Faculty of Transport and Traffic Sciences, University of Zagreb, Vukelićeva 4, 10000 Zagreb, Croatia

T. Mihetec
Croatia Control Ltd., Rudolfa Fizira 2, 10410 Velika Gorica, Croatia

I. Štimac
Zagreb Airport Ltd., Rudolfa Fizira 1, 10410 Velika Gorica, Croatia

M. Petrović et al. (eds.), *The Science and Development of Transport—ZIRP 2021*,
https://doi.org/10.1007/978-3-030-97528-9_12

the future needs and possibilities of the airport with regard to current capacities and existing infrastructure, in terms of increasing business interest in air cargo transport.

Keywords Intermodality · Economic growth · Cargo city · Logistic

1 Introduction

Airport cities and airports are valuable metropolitan and national assets. Properly planned and developed, they promote local and regional business and trade development attracting additional passengers and cargo. The integration of Franjo Tuđman Airport in Zagreb, Croatia, into a single European transport system, based on the principle of intermodality and sustainable development, would make the airport attractive in terms of a regional cargo centre for the countries of Southeast Europe. The availability of an efficient access network, including railways, motorways and roads in the urban and gravitational area of the airport, is an essential prerequisite for the development and competitiveness of the airport. From the aspect of geo-traffic and socio-economic factors, the importance of positioning the airport on the main transport corridors and the proximity of developed industrial centres will significantly affect its strategic and sustainable development. Franjo Tuđman Airport is the largest airport in the Republic of Croatia and one of the largest airports in the region and has the lowest seasonality compared to other international airports in the Republic of Croatia. Such predispositions enable even year-round traffic and connection with destinations outside the Republic of Croatia, and it is positioned as a desirable destination outside the tourist season. The concession agreement for the construction and management of Franjo Tuđman Airport, between the Republic of Croatia and Zagreb International Airport d.d. includes the Franjo Tuđman Airport Development Plan through the construction of a new passenger terminal and ancillary infrastructure, using a public–private partnership model over a period of 30 years, starting in December 2013. In addition to the development of the passenger terminal, outside the concession zone, an area has been reserved for further development and expansion of the airport and its complementary facilities of Airport City and Cargo City. The construction of modern and functional infrastructure capacities in the function of air freight transport are the basic prerequisites for the successful implementation of the project, as well as the transformation of the airport into the centre of metropolis economic activity. Global air traffic growth forecasts show significant growth in passenger traffic in the period 2017–2036. In the same period, the growth of freight transport will triple [1]. It is evident that increasing the number of passengers and cargo at the airport will strengthen the economic position of the local economy, attract investment, create new jobs, increase trade and improve the operational efficiencies of companies and areas around the airport [2]. The results of the study show that the airport is one of the largest employers and creator of additional value in the country, as shown by examples of the correlation of analysed direct and indirect economic impacts on employment, production and taxes in the region [3]. This paper's methodology and terminology contribute to further research of the Franjo Tuđman airport development possibilities as the future cargo city regional centre. The paper also provides guidelines to be applied in the analysis of the economic impact of Franjo Tuđman Airport on the local economy and the development of air freight transport. All above-mentioned enables better planning and the possibility of expanding of the airport capacities. Also, it provides better availability of public transport in order to increase mutual integration, which would improve regional connectivity with urban agglomerations.

The paper consists of five chapters. The second chapter summarizes the main objectives of the European Transport Strategy with an emphasis on intermodality. The third chapter analyses the impact of air cargo on the global economy through the growth of economic activities, employability and GDP. The fourth chapter analyses the current state of freight transport at the Franjo Tuđman Airport, which was taken as a case study for the purposes of this paper. The chapter also presents concrete proposals for increasing freight traffic at the airport, which should be fully realized with the construction of Cargo City.

2 European Transport Policies and Multimodality

The European Union (EU) wants to build a modern integrated transport system that strengthens the EU's global competitiveness, which can meet the challenges of sustainable, smart and inclusive growth. The first step towards these goals is to provide a functional infrastructure that can transport people and goods efficiently, safely and sustainably. In 2017, the EU's infrastructure included over 217,000 km of railways, 77,000 km of motorways, 42,000 km of inland waterways, 329 key seaports and 325 airports [4]. Intermodality, an integrated approach between transport systems such as rail and air transport is high on the political agenda in Europe. European airports should improve their multimodal connections to increase the efficiency of the transport network and passenger mobility.

2.1 Regulatory Framework for the Development of Trans European Network-Transport (TEN-T)

The implementation of the objectives and standards set in the guidelines for the development of the trans-European transport network (Regulation (EU) No 1315/2013 [5], TEN-T Regulation) from 2013 is crucial for a more sustainable and smarter European transport network. TEN-T policy deals with implementation and development of the European network of railways, roads, inland and maritime waterways, ports, airports and railway terminals. The TEN-T guidelines establish a Comprehensive Network and a Core Network that promotes better accessibility of all regions to global markets and places a strong focus on the infrastructure of strategic importance. The core and comprehensive network focus on modal integration, interoperability and coordinated infrastructure development, especially on cross-border sections to remove bottlenecks. The TEN-T regulation, thus, sets clear deadlines for the completion of the core network by 2030 and the comprehensive network by 2050 [6, 7]. Furthermore, significant investments in TEN-T are made through EU support from the Cohesion Fund and the European Regional Development Fund, as well as through various financial instruments provided by the European Investment Bank. In search for deeper and closer economic and political integration, the European Union has developed a number of tools that have enabled it to achieve its goals, and the author of [8] divides them into a regulatory and infrastructural framework. Although the two types of tools complement each other, the European Union has realized that infrastructures can be of great help in advancing market integration and helping those less developed Member States to achieve these goals. Based on the Commission's progress report on the implementation of the Trans-European Transport Network (TEN-T), it can be concluded that significant progress was made during 2016 and 2017 (reporting period), in terms of both technical compliance and financial investment. The largest share of total investments (€ 80 billion) reported by Member States (including part of EU co-financing wherever needed) is invested in the core network (€ 56.5 billion) to fill compliance gaps [6].

The Republic of Croatia is located on two corridors of the Basic Transport Network, on the Mediterranean Corridor and on the Rhine-Danube Corridor. The Mediterranean Corridor connects ports in the south-western Mediterranean area with the centre of the EU, following the coasts of Spain, France and crossing the Alps to the east. It passes through northern Italy and continues east, through Slovenia, Croatia and Hungary to the Ukrainian border [9]. It is a road and railway corridor, and it is an integral part is the route Rijeka–Zagreb–Budapest (Pan-European Corridor Vb). The Zagreb–Slovenia road and rail route (Pan-European Corridor X) continues to the Mediterranean Corridor. The Rhine–Danube corridor is an intermodal route connecting Strasbourg, Frankfurt, Vienna, Bratislava, Budapest, from where one part goes to Romania and the other goes along the Danube between the Republic of Croatia and Serbia and further to the Black Sea (Pan-European Corridor VII) [10]. In the context of this paper, the Mediterranean Corridor is interesting, as a multimodal corridor that enables the integration of the national road network into a single TEN-T network, which enables the development of freight traffic at the Franjo Tuđman Airport in the full sense.

The Mediterranean corridor includes 17 main airports: six are located in Spain (Valencia, Alicante, Seville, Malaga, Barcelona, Madrid—Barajas); two in France (Lyon Saint-Exupery and Marseille-Provence); six in Italy (Bergamo-Orio al Serio, Milan—Malpensa, Milan—Linate, Venezia—Tessera, Turin-Caselle, Bologna-Borgo Panigale) and one in the capitals of Slovenia, Croatia and Hungary. According to EU guidelines, only airports that have direct rail services that connect the airport with high-speed or TEN-T long-distance rail lines will be considered properly "connected to the rail". Local or regional/suburban rail links, although improving accessibility, are not sufficient to fully comply with the Regulation.

2.2 Intermodality as a Solution for Better Connectivity and Mobility

The Aviation Strategy for Europe [11] indicated that better air connectivity of a city, region or country with other destinations in Europe and other parts of the world could lead to greater growth. The availability of highly efficient and competitive services at airports, including runways, passenger terminals and ground handling services, is key to the competitiveness of the EU aviation sector and passenger satisfaction with the quality of services. A European Commission study on the impact of TEN-T completion on growth, jobs and the environment shows that improvements in travel times on rail and inland waterways will lead to a modal shift that will attract passengers and cargo for these modes of transport. The study also indicates that these strong improvements in rail freight performance reflect the impact of projects aimed at making long-distance rail freight more competitive with road transport [12]. Further progress is expected in the development of transport infrastructure, as the deadlines for the implementation of TEN-T approach (2030 for core network and 2050 for comprehensive network). Regarding airports, the TEN-T Regulation defines that only key airports in Annex 2 to Regulation no. 1315/2013 (i.e. 38) are subject to the obligation set out in Article 41. Furthermore, airports must be connected to the rail and road transport infrastructure of the trans-European transport network by 2050 (except where physical constraints prevent such a connection) and integrated into the high-speed rail network wherever possible, considering potential traffic.

The Air Transport Action Group (ATAG) promotes the intermodality of air and rail transport in all regions where it is possible. For example, in Europe, ATAG has contributed to a detailed study on identifying barriers and developing solutions to promote air and rail intermodality in the European Union. Intermodality is presented as a solution to many transport problems faced by modern societies (e.g. rising accident levels, CO_2 emissions and transport noise) and plays an important role in improving passenger mobility. A partial or complete replacement for air travel can be successful on short or medium journeys lasting up to three hours provided by a high-speed train (e.g. between Brussels and Paris). This requires that train timetables and flight plan, fares and handling activities are carefully coordinated. The next step will require an "integrated approach" with a synchronized information and distribution system in the aviation and rail system. Therefore, for airport operators, construction and real estate management projects are becoming an important business segment, often creating higher profit margins than the primary function of the airport, which is to provide infrastructure and services to airlines [13].

3 Positive Effects of Cargo City Infrastructure on Airport Development and Regional Economic Growth

Cargo carried by air represents less than 1% of global trade in tonnage, but air transport carries more than $6 trillion in goods each year, representing more than 35% of global trade in value [14]. Such a large difference between tonnage and value reflects the unique position of air transport in transporting goods, which often requires a high level of speed, reliability and safety. Time-sensitive products, computer equipment, machinery and electrical equipment are widely represented in air transport and account for the largest share of tonnage in air transport in relation to the tonnage of containers. European airports contribute to the employment of 12.3 million people, generating € 356 billion in revenue annually and generating € 675 billion in gross domestic product (GDP) each year, which is 4.1% of Europe's GDP. The activities of airport operators, air carriers, air traffic control, general aviation, airport equipment operators, airport security, immigration and customs, aircraft maintenance and other airport-related activities result in almost 1.7 million direct jobs in Europe [15].

A survey conducted by ATAG shows that 52% of companies consider international transport links as a key factor in locating their companies in Europe. The lack of good air transport connections significantly influenced companies' investment decisions [16]. Also, rapid commercial development of major airports (and around major airports) promotes them as the leading generator of urban growth. Airport areas are developing a "Brand Image" that affects even non-airport-related businesses [17].

The impact of the UK leaving the European Union on 31 January 2020, is affecting air freight and mail transport quantities at EU level. After several years of growth, in 2019 compared to 2018, there was a decline in total freight traffic in the EU (-4.1%). Between 2018 and 2019, domestic freight and mail transport recorded an increase of 0.5%, while intra-EU freight and mail transport reported a decrease of 5.3%. The dynamics of freight and mail transport by air between 2018 and 2019 differ significantly at the individual Member States level, with growth rates ranging from -30.9% in Malta to +14.4% in Finland [18]. In the EU-27, Paris/Charles de Gaulle has become a major European airport in terms of loading/unloading cargo and mail. Frankfurt Main has become Europe's second-largest cargo and mail airport after a year at the top.

Among the top 20 airports, 14 airports recorded a decline in the transport of goods between 2018 and 2019 in terms of total cargo and mail (loaded/unloaded). Amsterdam Schiphol recorded the largest decline in transported cargo and mail (−7.9%) and a decline in the number of cargo flights of 12.0%. Helsinki Vantaa shows a growth of 15.1% in terms of loaded/unloaded cargo and mail between 2018 and 2019, as shown in Table 1 [18].

The economic analysis of the impact of airports on regional development, commissioned by ACI Europe, shows the results of a multivariate analysis of the impact of air travel on the local economy and the impact of air connectivity on GDP and employment within a radius around the airport. Economic growth stimulates passenger

Table 1 Top 20 airports in the EU-27 in terms of total freight and mail loaded/unloaded in 2019 (t) [18]

Rank	Country	Airport	Total air transport (in tonnes)	Freight and mail loaded	Freight and mail unloaded	Growth of total freight transport 2018–2019 (%)	Total number of freight flights (in 1000)	Growth total number of freight flights 2018–2019 (%)
	FR	PARIS/CHARLES DE GAULLE	2.102.018	1.114.734	987.284	−1	27	−1.5
2	DE	FRANKFURT/MAIN	2.089.281	1.065.980	1.023.302	−4	22	1.5
3	NL	AMSTERDAM/SCHIPHOL	1.592.205	791.712	800.493	−7.9	13	−12
4	DE	LEIPZIG/HALLE	1.227.256	640.165	587.09	1.5	46	0.5
5	LU	LUXEMBOURG	853.028	425.242	427.786	−4.7	10	−1.5
6	BE	LIEGE	808.545	398.06	410.485	1.1	26	12.8
7	DE	KOLN/BONN	799.1	392.241	406.859	−5.4	33	6.1
8	BE	BRUSSELS	561.532	264.244	297.288	−4.3	14	−3.3
9	IT	MJLANO/MALPENSA	558.077	325.087	232.99	−3.4	14	13.2
10	ES	MAORID/BARAJAS	511.167	262.911	248.256	3.9	12	6.9
11	OE	MUNCHEN	349.958	201.31	148.648	−4.9	4	−5.8
12	DK	KOBENHAVN/KASTRUP	226.161	135.854	90.307	0.7	6	5.7
13	FI	HELSINKI/VANTM	221.452	121.201	100.245	15.1	3	23.9
14	AT	WIEN/SCHWECHAT	220.831	101.329	119.502	−3.8	5	3.7
15	1 T	ROMA/FIUMICINO	194.508	119.581	74.927	−5.5	2	47.6
16	PT	LISBOA	154.319	81.154	73.165	14.2	3	−4.7
17	ES	BARCELONA/El PRAT	143.143	74.459	68.684	−4.1	4	−1.7
18	DE	FRANKFURT-HAHN	142.414	77.539	64.876	−4.8	3	−1.9
19	IE	DUBLIN	133.229	61.871	71.358	−7.3	4	−1.7
20	IT	BERGAMO/ORIO AL SERIO	119.041	61.711	57.33	−2.4	8	−1.3

demand in air transport, which drives air connectivity, while increasing air connectivity again leads to an improved business climate and attracts tourism, which further boosts economic growth [19]. The authors [20] state that the need for quick transportation of standard semi-finished products for production is increasing. In this case, air transport plays a significant role. With the construction of adequate infrastructure, Cargo City strongly affects the overall development of the airport and the areas that gravitate to the airport. Without adequate infrastructure, significant development of the airport and surrounding area is not possible. Infrastructure should be planned modularly, considering traffic and economic growth projections.

4 Overview of the Current Situation and Guidelines for the Development of Freight Traffic at Franjo Tuđman Airport

Franjo Tuđman Airport is located at the intersection of the main road and railway corridors which, in synergy with the air corridors, create all the preconditions for developing a quality intermodal transport system. Connecting the basic network of transport infrastructure with trans-European networks and corridors is one of the basic goals and an important precondition for the balanced transport development of the Republic of Croatia. Therefore, the EU is continuously making efforts to build the necessary roads and integrate the national road network into a single TEN-T network. The existing motorway network (shown in Fig. 1) near Franjo Tuđman Airport is well developed and provides good connections within the Republic of Croatia. National motorways and international routes have intersection at the Zagreb: A1 Zagreb—Split, A2 Zagreb—Macelj, A3 Bregana—Zagreb—Lipovac, A4 Zagreb—Goričan, A6 Zagreb—Rijeka and A11 Zagreb—Sisak (under construction) [21, 22]. The main connection of the city of Zagreb with the Franjo Tuđman Airport is the state road D30 and the Homeland Bridge on the east side of the airport. The A11 motorway provides access to the wider gravitational area from the south via Velika Gorica junction south. Franjo Tuđman Airport provides access not only to the city of Zagreb, but also to other cities located in the region of Central Croatia (functional urban area consisting of the city and its travel zone whose labour market is highly integrated with the city) which include: Velika Gorica ($\approx$3 km), Varaždin ($\approx$87 km), Čakovec ($\approx$106 km), Koprivnica ($\approx$98 km), Bjelovar ($\approx$88 km), Virovitica ($\approx$153 km), Daruvar ($\approx$129 km), Zabok ($\approx$51 km), Zaprešić ($\approx$33 km), Kutina ($\approx$83 km), Sisak ($\approx$46 km) and Karlovac ($\approx$59 km). The connection between these cities and the Franjo Tuđman Airport is realized through several motorways (A1, A2, A3, A4, A11) and state roads with Zagreb as a starting point, or destination.

The Zagreb railway junction (Fig. 3) is connected to the trans-European transport network (TEN-T) via corridors RH1 and RH2 and represents the central core of the railway network in the Republic of Croatia, which connects the south-western and southern coastal network with the northern and eastern continental parts, with the

Fig. 1 Motorway network in the Republic of Croatia, Source. Adapted by the author according to the source [21, 22]

European transport system. The city of Zagreb, and, thus, the Zagreb railway junction, is located at the traffic junction of roads between Western, Central and South-eastern Europe and the Adriatic Sea [23]. Due to the position of the railway infrastructure and the network of motorways in the immediate vicinity, Franjo Tuđman Airport is positioned as an entity of importance for the development of complementary facilities, especially Cargo City.

In terms of competition, there are several airports where cargo could be redirected to the Zagreb International Airport vicinity, and those airports could be defined as Zagreb Airport competitors. Those airports are Belgrade, Ljubljana, Sarajevo and Venezia Marko Polo. Although many studies mention Vienna, Budapest and Bergamo airports as competition to Zagreb Airport due to their high capacity infrastructure, air cargo network and level of technology, it is not realistic that Zagreb Airport could compete with those airports. However, it can be positioned as s small regional cargo hub and compete with Ljubljana, Belgrade, Sarajevo and Venice. In the Fig. 2, cargo traffic in those airports can be seen.

Related to completion from Ljubljana Airport, it can be said that this airport is weaker in cargo handling, especially from the time that Adria Airways (Slovenian national flag carrier) got bankrupt and that DHL repositions its aircraft to Zagreb. At Sarajevo Airport, the problem is similar because there is no national carrier that can significantly increase cargo transport. Belgrade Airport and Venice Airport are much stronger competitors for Zagreb. Belgrade Airport has a good infrastructure, and it is base for the Serbian national carrier who is in partnership with one of the world's greatest airlines, Etihad. Their collaboration in networking (intra-Europe and international long-haul) place Belgrade Airport on the map as a strong passenger and

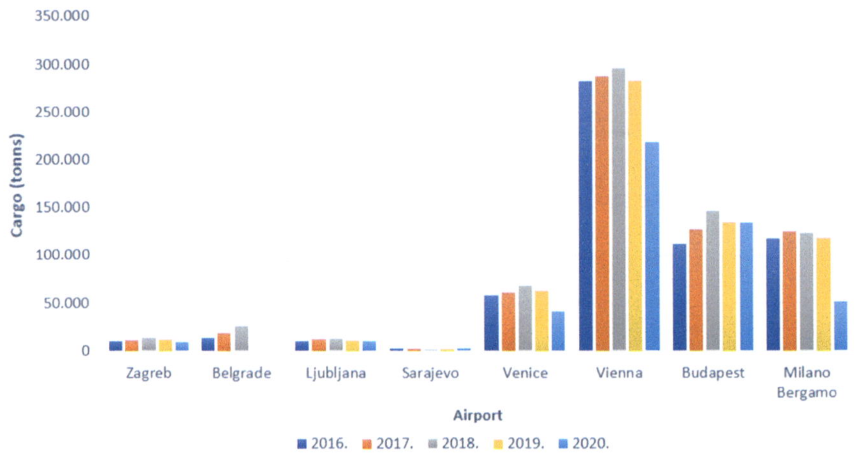

Fig. 2 Air freight statistics for Zagreb Airport and its competitors [24]

Fig. 3 Zagreb railway junction, Source: Adapted by the author according to the source [23]

cargo hub. Regarding Venice Airport, one of the main road corridors regarding cargo is passing just a few kilometres from Venice Airport, and it gives a huge advantage for developing intermodality and cargo hub at Venice Airport.

Knowing abovementioned facts, Zagreb Airport can become regional passenger and cargo hub only with huge investments, quality and efficient management and modern infrastructure. Due to previously mentioned, the Government of the Republic

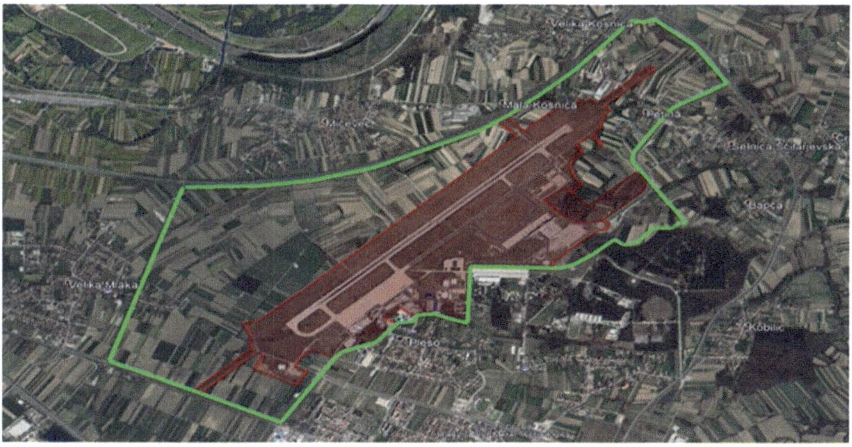

Fig. 4 Area of concession (red area) and maximum development (green line), Source: Created on Google Earth Pro

of Croatia decided to put Zagreb Airport under concession based on the BOT (build-Operate-Transfer) model. Following that, since December 2013, Franjo Tuđman Airport has been granted a concession for a period of 30 years. The subject area of the concession is shown in red in Fig. 4 and includes the airport infrastructure with a new passenger terminal, as a hub facility, which was built and opened in 2017. Figure 4 shows that the area of maximum expansion of Franjo Tuđman Airport is much wider than the area given in the concession (green line), and the same is reserved in the spatial plans of the city of Velika Gorica and Zagreb County.

The area outside the concession zone is reserved for complementary facilities at the airport, namely the Airport City and Cargo City projects. Airport City and Cargo City are separate projects; however, the development of the Airport city, of which Cargo City is an integral part, proved to be far from simple. Airport City is a project that focuses as complementary content to passengers and business users of the airport. Construction of Airport City is planned at the east side of the new passenger terminal and will include, for example, hotel, office buildings, catering, entertainment and other related facilities [25]. Cargo City, a project whose implementation, is planned in the north-western part of the airport, creating triangle area consisting of the Zagreb motorway, the railway from the marshalling yard train station and the airport runway. The development of infrastructure through the implementation of the Zagreb Cargo City project implies the construction of a modern cargo infrastructure which consists of storage space; office space; runway; apron for cargo aircraft; taxiway system; truck and car parking lots; access roads; the industrial track of the railway and other facilities as shown in Fig. 5 [26].

In order to position the Franjo Tuđman Airport as a competitive regional cargo centre, it is necessary to invest significant funds in infrastructure and superstructure. It is necessary to build a high-tech fully computerized complex centralized system

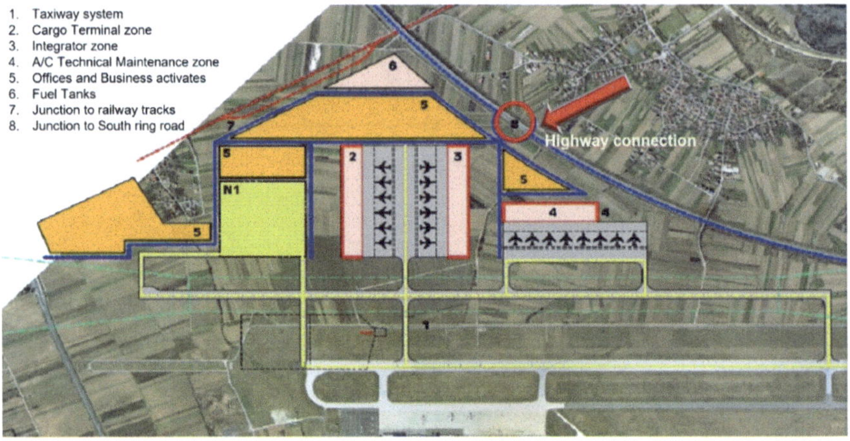

Fig. 5 Cargo City development area, Source: Zagreb Airport Masterplan [26]

of terminals (minimum semi-automatic, preferably fully automatic) that would be able to accept and ship special and dangerous goods, build a railway line that will connect the main railway line M502 Zagreb (Central Station)—Sisak-Novska and the Franjo Tuđman Airport zone (including Zagreb Airport Cargo City) and to build road connection of Zagreb Airport Cargo City to the A3 motorway (European route E65). From the aspect of suprastructure, in order to be competitive, it is necessary to modernize the existing capacities of the cargo terminal (systems for consolidation and deconsolidation of shipments, systems for palletizing shipments, implement new X-ray devices in accordance with the latest standards), as well as introduce radio-frequency identification (RFID) technology in order to increase the level of quality of service. Analysing the quantities of transported cargo and mail at the Franjo Tuđman Airport, regardless of the adequacy of capacity for receiving and dispatching goods, there has not been a significant increase in freight traffic last 20 years. Freight traffic growth is slower than in the 1990s, when freight traffic exceeded 16,000 tons, as shown in Fig. 6.

Statistics over the last six years show an optimistic trend in terms of recovery and a slight increase in air freight transport (Fig. 7). In 2020, freight traffic increased compared to 2019 by 5.84%, mostly due to the need for air freight to mitigate the effects of COVID 19, while mail transport had a decline of as much as 45.26% compared to the same period. Compared to other airports in the Republic of Croatia (six international airports), almost 90% of all cargo is accepted and shipped through Franjo Tuđman Airport [28].

World air freight traffic is projected to grow 4.2% per year over the next 20 years. In terms of Revenue Tonne-Kilometres (RTK) growth, air traffic growth, including express traffic, is projected at a rate of 4.3 percent per year, while airmail will grow slowly, averaging 2% per year until 2037. Risks that could affect the future growth of airmail include increasing reliance on Internet communication and stricter security requirements. Overall, global air freight traffic will more than double in the next

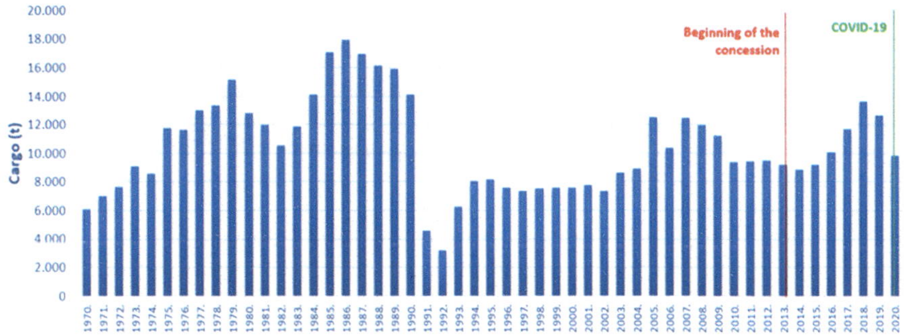

Fig. 6 Cargo transported at Franjo Tuđman Airport in the period from 1970 to 2020 [27]

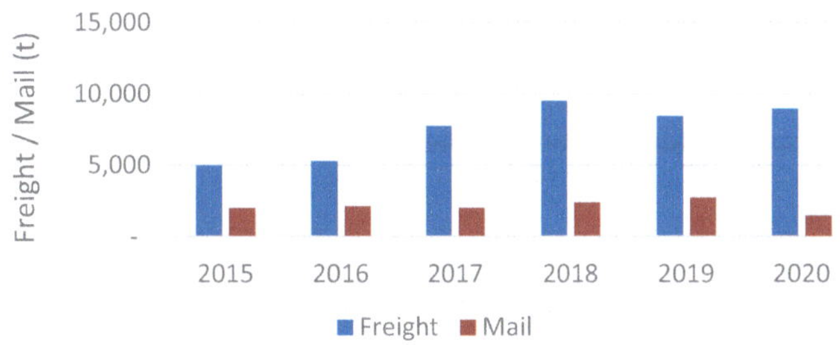

Fig. 7 Franjo Tuđman Airport cargo traffic for the period 2015–2020 [28]

20 years, increasing from 256 billion RTK in 2017 to 584 billion RTK in 2037 [29]. According to available traffic forecasts for Europe, the growth rate of air cargo from 2020 to 2039 is estimated at an annual rate of 2.3% [30]. This forecast is considered a sufficient reason to take action to continuously improve the quality of the process of receiving and dispatching cargo at the Franjo Tuđman Airport, as well as investing in infrastructure and designing models for future business. Taking into account the growth rates of freight transport at the European level of 2.3%, the long-term forecast of air freight and mail services demand for the period from 2020 to 2039 at the Franjo Tuđman Airport predicts an increase in demand in domestic and international turnover from 8,000 tons in 2020 to about 15,000 tons in 2039, or about 57% increase. If expectations were met in line with global air cargo growth (4.2%), freight figures would reach 20,000 tonnes per year (shown in Fig. 8). The flow of goods (annual volumes, frequencies, characteristic goods, etc.) that can potentially be served as air cargo within the airport's gravity area is the only realistic indicator for Franjo Tuđman Airport's positioning in the air cargo market, through the construction of the necessary infrastructure for receiving and dispatching cargo (Cargo City). By promoting compatibility between modes of transport for the reception and dispatch of express mail and cargo, the Cargo City project will also promote the economic growth of air transport and thus support the efforts of a sustainable development strategy. Given the long-term projection of air cargo traffic, it is necessary to devise

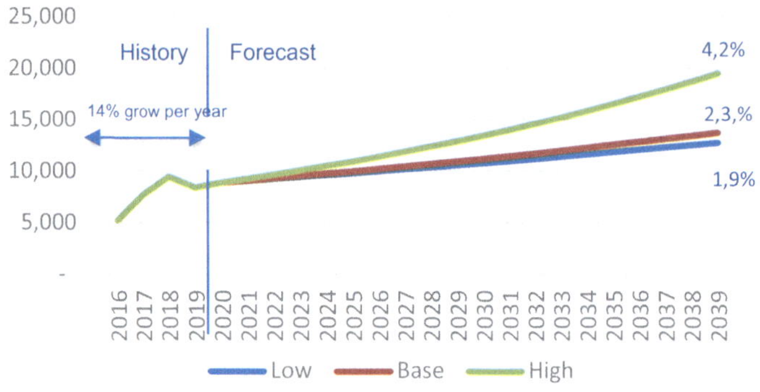

Fig. 8 Cargo traffic average annual growth projection for the period 2020–2039. *Source* Adapted by the author based on statistics

a business model and assumptions for the positioning of the Republic of Croatia, respectively, Franjo Tuđman Airport on the market of cargo traffic (Fig. 8).

The new business model should be viewed in the context of global and regional factors that will dominantly affect cargo traffic based on activities that are developing in three directions: (1) technology development; (2) infrastructure development and (3) traffic development [31]. Air freight transport can offer regional airports an additional opportunity on top of passenger transport segment, making them more profitable and attractive to regional business structures. Therefore, it is necessary to assess the impact of investment in transport infrastructure, particularly on competitiveness and economic growth, and to analyse current practices in the approach to assessing investment and direct effects in air freight markets. Infrastructural transport connections between transport modes are of great importance for the airport development, especially connecting them with the railway system, directly connecting the airport with the main railway networks in the EU. A rough assessment could be made for the development of the cargo centre as a regional hub, after analysing the parameters as mentioned above, i.e. current capacities of the airport and its facilities, future needs and expansion plans. Such an assessment would present the basis for the strategic settings for the developing of Franjo Tuđman Airport and Cargo City.

5 Conclusion

Forecasting long-term demand for air transport services requires assumptions and forecasts about macroeconomic trends in the air transport market. Many factors often vary from market to market. However, three key dimensions of the macro-environment that drive demand forecasting can be categorized as follows: the basic demand for air travel, the regulatory, infrastructural and technological environment, and the strategies and products that airlines offer in the market. Upgrading airports and expanding routes are vital for cities and regions looking to diversify their economies, increase exports and attract investment and tourists with high solvency.

For potential investors and tourists coming by aircraft, airports are a reminder of the metropolitan region and the first and last impression of the region. The availability of highly competitive airport services, including runways, passenger terminals and ground services, as well as connectivity to trans-European networks and corridors, is crucial to the competitiveness of Franjo Tuđman Airport. The specificity of the geo-traffic position of Franjo Tuđman Airport in the function of increasing regional competitiveness is primarily reflected in the expansion of airport capacity and the ability to provide services to airlines, which compared to competing airports in the area, represents an efficient and secure system for passengers, cargo and mail handling. Airports have become the primary factor of urban economic growth—while at the same time developing their areas in collaboration with companies such as shopping clusters, hotels, office complexes and conference and exhibition facilities, on the other side are attracting a concentration of business activities previously limited to urban centre. Air transport services increase economic progress in developing countries as they connect markets across the continent.

An increase in freight traffic may require considerable effort from airports, given the complexity of the project and the potential investment required. It is therefore important to study its potential in attracting cargo flows and devise a conscious freight transport development strategy, which includes local communities and the Government, through a quantitative and qualitative analysis of the economic impact of air cargo transport. However, to support an effective link between urban logistics and airport-oriented development, the economic growth of the wider gravitational area should have significant economic potential. This paper provides answers in which direction it is necessary to implement infrastructure capacity development policies in order to position the airport as a regional hub for freight transport.

The obtained results indicate the importance of integrated management and planning in the entire urban region, which has not been fully implemented so far, including the airport, by providing an efficient transport network and modern infrastructure for receiving and dispatching cargo and mail.

References

1. IATA & Airports Council International (2018); Boeing current market outlook 2017–2036)
2. Kasarda JD (2019) Kenan-Flagler Business School, University of North Carolina—Chapel Hill-AAAE Workshop, Indianapolis, IN, 06 May 2019.
3. Dusek T, Lukovics M, Bohl P (2011) The economic impact of the Budapest Airport on the local economy. European Regional Science Association
4. European Commission (2021). Mobility and Transport, from European Commission, https://ec.europa.eu/transport/themes/infrastructure_en. Last Accessed 27 Mar 2021
5. Regulation (EU) No 1315/2013 of the European Parliament and of the Council of 11 December 2013 on Union guidelines for the development of the trans-European transport network and repealing Decision No 661/2010/EU, (OJ L 348, 20.12.2013, p. 1)

6. COM (2020) 433 final—Report from the Commission to the European Parliament, the Council, the European economic and social committee and the Committee of the regions, Progress report on implementation of the TEN-T network in 2016–2017. Brussels, 26 Aug 2020
7. Regulation (EU) No 1316/2013 of the European Parliament and of the Council of 11 December 2013 establishing the Connecting Europe Facility, amending Regulation (EU) No 913/2010 and repealing Regulations (EC) No 680/2007 and (EC) No 67/2010 Text with EEA relevance
8. Rodrigo, B. Prieto.: Common European Transport Infrastructure Policies – Do they increase economic growth and improve market efficiency? University of Alcalá, Alcalá de Henares, Madrid, Spain, Published: April 2018
9. EC Mobility and Transport, Mediterranean Fourth Work Plan of the European Coordinator, Iveta Radičová, August 2020
10. Ministarstvo mora, prometa i infrastrukture, EU prometni koridori i TEN-T. https://promet-eufondovi.hr/eu-prometni-koridori-i-ten-t/. Accessed 28 May 2021
11. European Commission (2015) An Aviation Strategy for Europe {SWD(2015) 261 final} Bruxelles, 7.12.2015.
12. European Commission (2019) Mobility and Transport, Studies on Transport issues-The impact of TEN-T completion on growth, jobs and the environment
13. European Commission (2010) Topical Report - Airport Accessibility in Europe, September 2010
14. IATA. https://www.iata.org/cargo/. Accessed 14 June 2021
15. InterVISTAS (2015) Economic Impact of European Airports, A Critical Catalyst to Economic Growth—Prepared for ACI Europe January 2015
16. ATAG (2008) The economic & social benefits of air transport, Air Transport Action Group, Geneva , Switzerland
17. Sengupta PS (2007) Principles of airport economics, excel books, pp 179–206
18. Eurostat—Your key to european statistics. https://ec.europa.eu/eurostat. Accessed 27 Apr 2021
19. SEO AVIATION ECONOMICS (2015)-Regional economic impact of airports, Final report, Amsterdam, 20 October 2015, Commissioned by ACI EUROPE
20. Shepherd B, Shingal A, Raj A (2016) The value of Air Cargo: air transport and global value chains. Montreal: The International Air Transport Association (IATA)
21. Gradski ured za strategijsko planiranje i razvoj Grada, Razvojna strategija Grada Zagreba za razdoblje do 2020. godine, Zagreb, 2017
22. Croatian Motorways Ltd. Limited Liability Company. https://www.hac.hr/hr. Accessed 15 June 2021
23. Ministarstvo mora, prometa i infrastrukture, Studija razvoja željezničkog čvora Zagreb, Zagreb 2016
24. CAPA (2021) Centre for Aviation 2021
25. Drljača M, Petar S, Raad M, Štimac I (2020) The role and position of airport city in the supply chain. Prod Eng Arch 26
26. Drljača M, Kramarić G, Pašagić ŠJ (2012) Zagreb Airport Cargo City Project, ZIRP 2012, Faculty of Transport and Traffic Science Zagreb, Zagreb, pp 235–240
27. Zagreb Airport. http://zlz-zagreb-airport.hr/hr/statistika. Accessed 11 June 2021
28. Statistički podaci o prometu na zračnim lukama RH. https://www.ccaa.hr/statisticki-podaci-o-prometu-na-zracnim-lukama-rh. Accessed 31 May 2021
29. World air cargo forecast 2018–2037, Boeing Commercial Airplanes, Boeing, Commercial Airplanes P.O. Box 3707, MC 21–33, Seattle, WA 98124–2207 USA
30. Statista, Forecast of air cargo growth rates from 2020 to 2039. https://www.statista.com/statistics/1189158/air-cargo-growth-rates-by-region/. Accessed 14 June 2021
31. Drljača M (2012) Cargo traffic of Zagreb airport and business model. Suvremeni promet, Vol. 31, No. 1–2, Hrvatsko znanstveno društvo za promet, Zagreb, Opatija, 2012, str. 42–46

CFD Analysis of F-16 Wing Airfoil Aerodynamics in Supersonic Flow

Leo Matak and Karolina Krajček Nikolić

Abstract Aerodynamic characteristics of the NACA 64A-204 airfoil are evaluated by numerical analysis of the turbulent flow at compressible and high Mach number using computational fluid dynamics (CFD). The simulations were done using the Ansys Fluent 2021R Academic Version. The results are obtained with Reynolds averaged Navier–Stokes (RANS), and for resolving of the flow turbulence, the k-ω SST turbulence model was selected. The paper shows how a simple simulation-based analysis predicts better aerodynamical characteristics for a wing with changing geometry. It is shown how the maximum ratio of C_L/C_D for subsonic configuration is 6.9, whereas for supersonic configuration under equal conditions, the max value for C_L/C_D is 7.1, which is around 2.75% increase. Also, for supersonic flight regime, the supersonic airfoil configuration provides over 10% higher value for C_L/C_D ratio at 1° angle of attack based on the CFD results. This paper also promotes the wing morphing technologies due to increases in aerodynamical characteristics when the geometry changes. Changing wing geometry could lead to less fuel consumption and less noise creation above urban places with airports inside the city area. It is expected that wing changing geometry will be used not only for military but also for civil supersonic aircraft.

Keywords NACA 64A-204 · C_L/C_D · Variable camber · CFD analysis · F-16 wing airfoil

1 Introduction

This paper will focus on a study of the NACA 64A-204 airfoil used by the F-16 aircraft. The specified airfoil is selected because during the flight, the F-16 has an

L. Matak
Houston University, Calhoun Rd 4800, Houston, TX 77004, USA

K. K. Nikolić (✉)
Faculty of Transport and Traffic Sciences, Department of Aeronautics, Vukelićeva 4, 10000 Zagreb, Croatia
e-mail: kkrajcek@fpz.unizg.hr

M. Petrović et al. (eds.), *The Science and Development of Transport—ZIRP 2021*,
https://doi.org/10.1007/978-3-030-97528-9_13

ability to change the geometry of its wings, and thus its airfoil geometric characteristics, by deflecting the leading or trailing edge flaps or both. By doing so the wing effectively changes the curvature and the camber line of the airfoil. And since the curvature and camber line have a significant impact on the airfoil characteristics, it results in changes in lift, drag and pressure coefficients.

At supersonic speed, the F-16 automatic device schedule includes leading and trailing edges with a negative deflection, as shown in Fig. 1—the supersonic flight configuration (it will be denoted as configuration 1 from now on). The leading and trailing edges use an automatic −2° deflection at supersonic speeds in order to mitigate the negative lift. At subsonic cruise, the basic wing camber is unmodified, as in Fig. 1—the subsonic cruise configuration (it will be denoted as configuration 0 further in the paper).

Leading edge flap is automatically programmed for best flap position (max lift/drag as a function of Mach number and angle of attack). Variable camber shapes airfoil to match desired flight condition in order to maximize lift/drag ratio, improve directional stability and minimize buffet. There are additional configurations which are used by the F-16, but for the purpose of this analysis, only the two mentioned above are important. Table 1 shows different deflection behaviour of the leading and trailing edge flaps depending on the flight phase, along with the velocity/Mach number for each of the regimes.

Recently, there has been research conducted for various NACA 6 series airfoils, even for the NACA 64A-204. The research conducted in reference [1] shows the effects of relative thickness on the aerodynamic performance of the airfoils, by conducting similar numerical analysis in different software, however, only for much lower Mach values (Ma = 0.15). Akgun et al. [2] conducted drag calculation through CFD simulations using the Ansys software but for the entire F-16 aircraft. Hodson et al. [3] worked on a biologically inspired aeroelastic topology optimization method which they applied to a morphing airfoil in supersonic flow using evolutionary

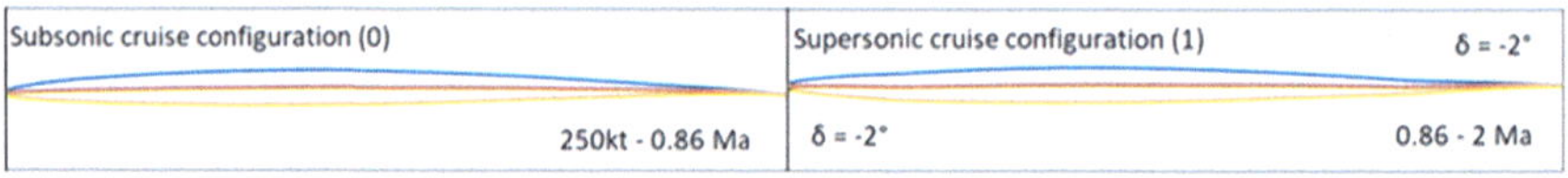

Fig. 1 Flaps' position configurations

Table 1 Flaps' deflection for different flight phases

Flight phase	Indicated airspeed/Mach number	Leading edge flaps' angle	Trailing edge flaps' angle
Ground roll	0–175 kts	−2°	20°
Take-off and landing	175–250 kts	15°	20°
Subsonic cruise	0.86 Ma	0°	0°
Supersonic cruise	0.86–2 Ma	−2°	−2°

design methods. Their research yields unique topologies that have provided valuable insights into the behaviour and capabilities of bio-inspired aircraft systems. Hari et al. [4] made a numerical analysis over a strut intersecting a wing configuration in a supersonic flow. They solved for the Navier–Stokes equations using the same solver as is used in this paper; however, they have based their research on the vertical angle made by strut to the wall. Askari et al. [5] worked on a similar research using the CFD method along with the Spalart–Allmaras turbulence model for computation of the Reynolds stresses; however, they based their simulation on the first-order upwind discretization. They concluded that the aerodynamic coefficients gained from both analytical and numerical methods were in good agreement. Keith et al. [6] used the shock-expansion theory for the computations of a spreadsheet for the well-known lift and drag characteristics of a symmetric double wedge airfoil in supersonic flow. Another interesting work for the optimization of the wing-design concept was conducted by Manshadi and Aghajanian [7]. They conducted a CFD analysis of the entire wing at supersonic flow conditions by using the response surface method. Since there was no easily obtainable literature available for the sole airfoil aerodynamical analysis for the lift and drag coefficients depending on the change of the geometry [8], this study will fill the missing data gap and can be used for data validation in other research studies.

2 CFD Modelling of Flow Past Airfoil

The numerical technique used to solve the governing equations of fluid flow is the finite volume method used by Ansys Fluent.

The first step of the CFD modelling is the geometry setup, for which the NACA 64A-204 coordinate data was computed in MATLAB and then imported and finalized in the Ansys Design Modeller—a CAD part of the Ansys software. Afterwards, mesh controls and setup were selected. Next, solver settings and solver preference were chosen, along with the boundary and flow conditions. The selected turbulence model for resolving the flow simulation was the k-ω SST model along with the pressure-based coupling algorithm. Finally, in the post-processing part, pressure and velocity contours were displayed. In Fig. 2, flowchart is given for better understanding of the process for obtaining the results.

2.1 *Geometry Setup*

For the geometry of the NACA 64A-204 airfoil, taken from Airfoil Tools, coordinate file was made and then exported from MATLAB into the Design Modeller.

Figure 3 shows how the airfoil looks in the MATLAB plotter before being exported for further modelling. The flow field was then created around the airfoil, and it was created in a C-section type. The radius of the half circle in front of the airfoil was 10

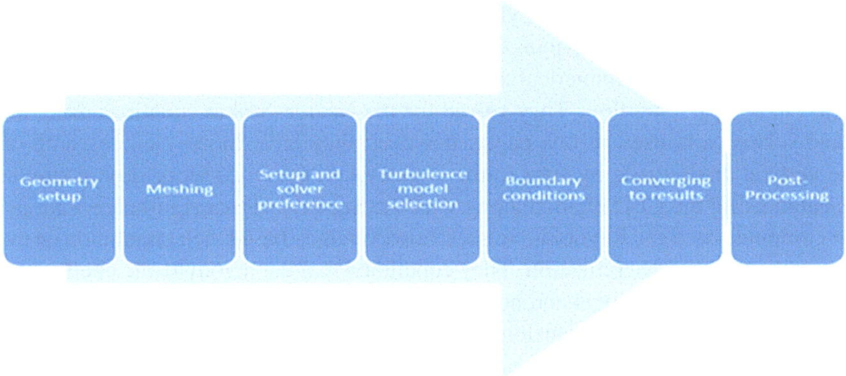

Fig. 2 Ansys Fluent flowchart for obtaining results

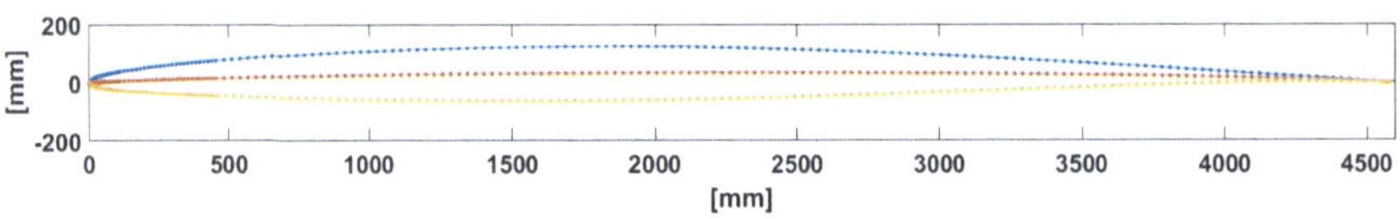

Fig. 3 Configuration 0 computed in MATLAB

chords long, whereas the length of the area behind the airfoil was 12 chords. Length of the airfoil chord is 4580 mm due to it being the length of the mean aerodynamical chord of the entire F-16 wing based on its wing root and wing tip lengths.

Figure 4 demonstrates how the flow layout looks in the design modeller of the Ansys software. Specified length values have been proven to be enough, as to not receive the reverse back flow, and other various effects which may occur if the flow field area is too small. The idea was to simulate the flow through the flow domain and to cut out the surface of an airfoil from the flow domain, so it acts as a wall over which the free-stream flows.

2.2 *Meshing*

After successfully creating the geometries of both the airfoil and the surrounding flow field, the next step in the process of obtaining the targeted results is discretization of the area or as it is better known—meshing. Meshing process is about 50% of the work.

After a successful transition of the model from the design modeller, meshing can be approached in two ways: global and local. Global mesh controls are applied to the entire grid, whereas local meshing options are available from the mesh control menu, and they are applied locally to selected features.

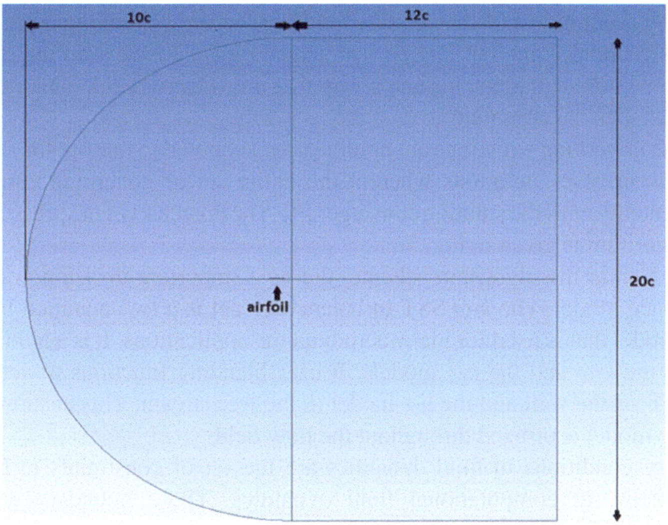

Fig. 4 CFD modelling of flow domain

By choosing the computational fluid dynamics for physics preference, option to select a preferred solver becomes available. For this study, the FLUENT solver is selected. Based on the selection, the software will try to generate the mesh which fits the solver.

2.3 Condition and Solver Option Set-up

After completing the mesh procedure, next step is setting up the conditions and preferred solver options. For this study, a pressure-based coupled algorithm is selected. Historically speaking, the pressure-based approach was developed for low-speed incompressible flows, while the density-based approach was mainly used for high-speed compressible flows.

However, recently both methods have been extended and reformulated to solve and operate for a wide range of flow conditions beyond their traditional or original intent [9]. In both methods, the velocity field is obtained from the momentum equations. In the pressure-based approach, the pressure field is extracted by solving a pressure or pressure correction equation which is obtained by manipulating continuity and momentum equations.

Using either method, Ansys Fluent will solve the governing integral equations for the conservation of mass and momentum, and (when appropriate) for energy and other scalars such as turbulence and turbulence dissipation rate. The pressure-based solver employs an algorithm which belongs to a general class of methods called the projection method. In the projection method, wherein the constraint of mass conservation (continuity) of the velocity field is achieved by solving a pressure (or pressure correction) equation. The pressure equation is derived from the continuity and the momentum equations in such a way that the velocity field, corrected by the pressure, satisfies the continuity.

Since the governing equations are nonlinear and coupled to one another, the solution process involves iterations wherein the entire set of governing equations is solved repeatedly until the solution converges [9]. The flowchart of the pressure-based coupled algorithm is given in Fig. 5.

After selecting the algorithm scheme, the next important decision is choosing the turbulence model. The *k-ω* SST turbulence model is a two-equation turbulent-viscosity model that is used for many aerodynamic applications. It is a hybrid model combining the *k-ω* and the *k-ε* models. It uses blending functions to activate the *k-ω* model near the wall and the k-ε model in the free stream. This ensures that the appropriate model is utilized throughout the flow field.

Boundary conditions in fluid dynamics are the set of constraints to boundary value problems in computational fluid dynamics. These boundary conditions include inlet boundary conditions, outlet boundary conditions, wall boundary conditions, constant pressure boundary conditions, axisymmetric boundary conditions, symmetric boundary conditions, and periodic or cyclic boundary conditions.

The governing equations are the same for whatever the flow is, whether it is over a large transport aircraft or past a windmill. However, the flow fields are quite

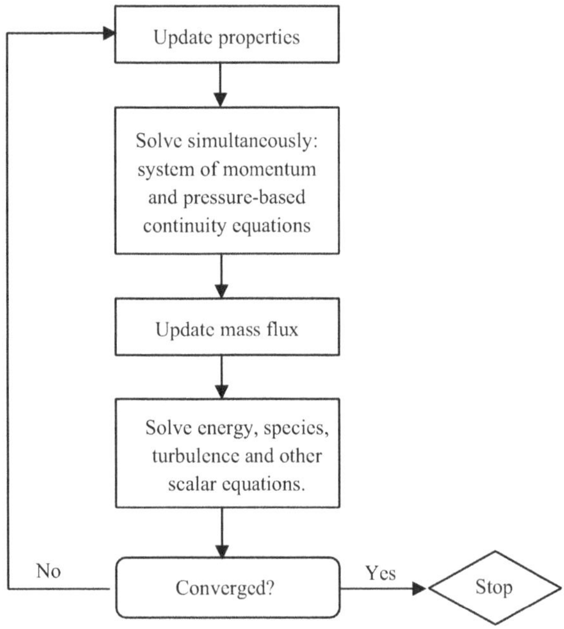

Fig. 5 Flow chart for obtaining CFD solution [9]

different for these cases, although the governing equations are the same. The difference enters with the boundary conditions which are quite different for each of the above mentioned.

Boundary conditions, sometimes called the initial conditions, dictate the particular solutions to be obtained from the governing equations.

Once the governing flow equations are obtained, then the real driver for any particular solution is the boundary conditions. This has a particular significance in CFD; any numerical solution of the governing flow equations must be made to see a strong and compelling numerical representation of the proper boundary conditions [10].

For a viscous flow, the boundary condition on a surface assumes zero relative velocity between the surface and the gas immediately at the surface. This is called the no-slip condition.

As for the inlet and outlet boundary conditions, particularly interesting is the pressure far-field condition. Pressure far-field conditions are used in Ansys Fluent to model a free-stream condition at infinity, with free-stream Mach number and static conditions being specified. The pressure far-field boundary condition is often called a characteristic boundary condition, since it uses characteristic information (Riemann invariants) to determine the flow variables at the boundaries.

2.4 Converging to Results

After specifying the flow and boundary conditions, the software goes from partial differential equations to algebraic equations and tries to iteratively guess the values for all the targeted variables. Ansys Fluent allows the user to specify the mass, momentum, energy, k and ω residuals for which to aim as to accept the results as correct. It is not always easy to achieve smooth convergence of the results. If there are low-quality mesh cells, or if the specified conditions are not physically possible, or if there are too large gradient changes in variable fields it is very easy for software to diverge and never to obtain proper results. That is why with the experience, the user learns how to guide the software in order to achieve the wanted convergence.

3 Results

For the simulation around the airfoil used by the F-16, a standard C-section type flow domain was created. It is a common approach in the external aerodynamical flows over airfoil surfaces. Grid is made in a structural way in order to obtain higher mesh element quality. Even though the finite volume method can work with irregular cell shapes, having a structured mesh helps in the amount of time required for the results to converge, as well as in the presentation of the results in the post-processing.

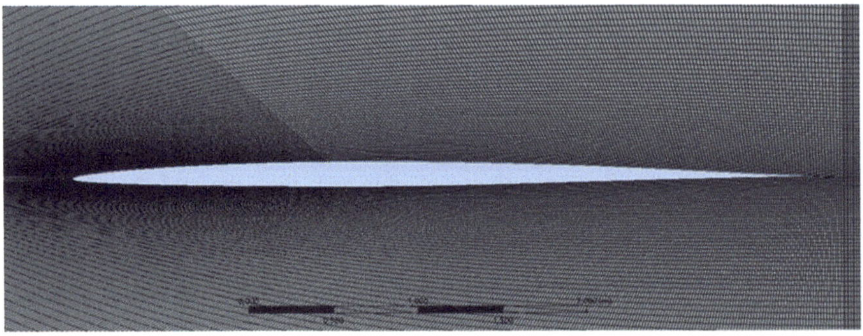

Fig. 6 Mesh around the airfoil

Table 2 Grid independence study at Ma = 1.5 and 5° AoA

Grid	Number of cells	C_L	C_D
#1	900	0.29	0.037
#2	7500	0.3	0.41
#3	41,250	0.31	0.44
#4	65,000	0.31	0.45

The entire grid domain is focused around the airfoil—which is the most important part of the flow field. Figure 6 shows the element distribution closely around the airfoil surface.

Grid independence study was conducted with different number of cells by considering the lift and drag coefficient values as parameters. Four different grids were evaluated with cell numbers of 900, 7500, 41,250 and 65,000 generated to evaluate grid independency. All the results are compared to each other. In order to calculate lift and drag coefficient Ma = 1.5 and 5°, angle of attacks (AoAs) are used. Results can be seen in Table 2. There is a negligible difference between the results of grid #3 and #4, and since grid #3 uses less cells to give equal results, grid #3 was selected in order to save computation time.

3.1 Setup Solver and Boundary Conditions

For the solver, the FLUENT solver along with the pressure-based algorithm is selected. Velocity formulation is absolute, and since no axes were used for any kind of symmetry, the 2D space is defined as planar. Time is steady—there are no changes to the flow after setting the initial boundary conditions. The point of the simulation was to conduct studies by solving full Navier–Stokes equations, and that is why the viscosity effects could not be omitted. Since that is the case, the turbulence model is set to viscous k-ω SST two equations model.

Material selected for the flow fluid is air, but in order to account for the compressibility at higher Ma numbers, the density is based on the ideal gas law. Viscosity of the air is based on the Sutherland's three coefficient law. Since the thermal reactions and thermal analysis are not the focus of this study, the material from which the airfoil is created is set as default and is not important [9].

For the boundary conditions, the flow inlet was selected as the far-field pressure inlet which is used to model free-stream compressible flow at infinity, with free-stream Mach number and static conditions specified. This boundary type is used only for compressible flows. As discussed earlier, the idea was to create a large enough flow field and then to cut out the surface of the airfoil so it will act as a wall over which the flow will stream. That is why the airfoil surface is defined as a no-slip wall boundary, which means that the velocity profile at the distance next to the wall is equal to 0.

The reference value used for calculation of the lift and drag forces and coefficients is based on the length of the airfoil chord. Since this is a 2D study case, the area of importance is simply the product of the airfoil chord length times one, $c = 4.58$ m, $A = 4.58\ \text{m}^2$.

3.2 Post-Processing

Figures 7 and 8 show the calculated pressure coefficient distribution along the length of the airfoil surfaces for configurations 0 and 1, respectively. The conditions of the simulation are Ma = 1.5 and 0° AoA. The difference between the black and red lines is what generates the lift and drag forces.

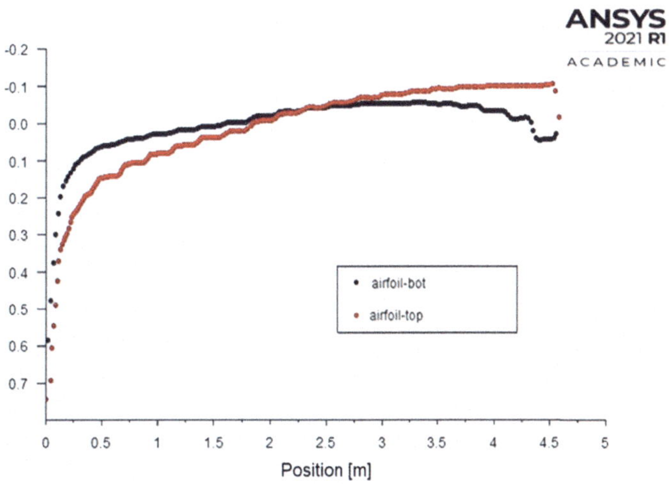

Fig. 7 CFD pressure coefficient distribution—configuration 0, Ma = 1.5, 0° AoA

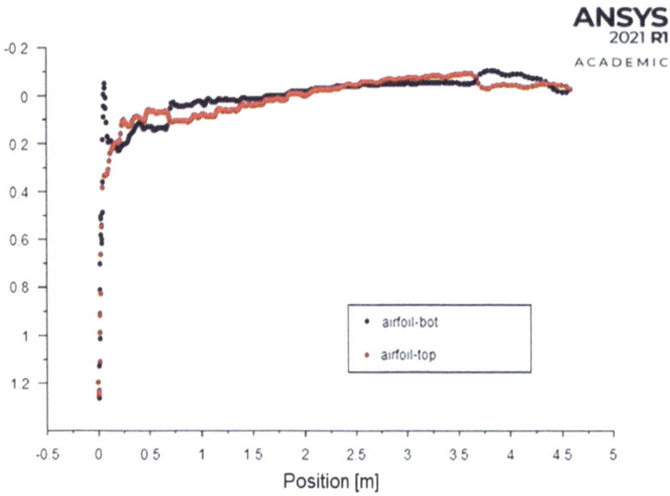

Fig. 8 CFD pressure coefficient distribution—configuration 1, Ma = 1.5, 0° AoA

From Fig. 7, it can easily be observed how there is a mild shockwave occurring at the rear edge of the airfoil which causes sudden changes in pressure gradient. It can also be seen how in supersonic flight regime it is not necessary to fulfil the Kutta condition at the rear end of the airfoil.

Figures 9 and 10 show the post-processing layout of the Ansys software.

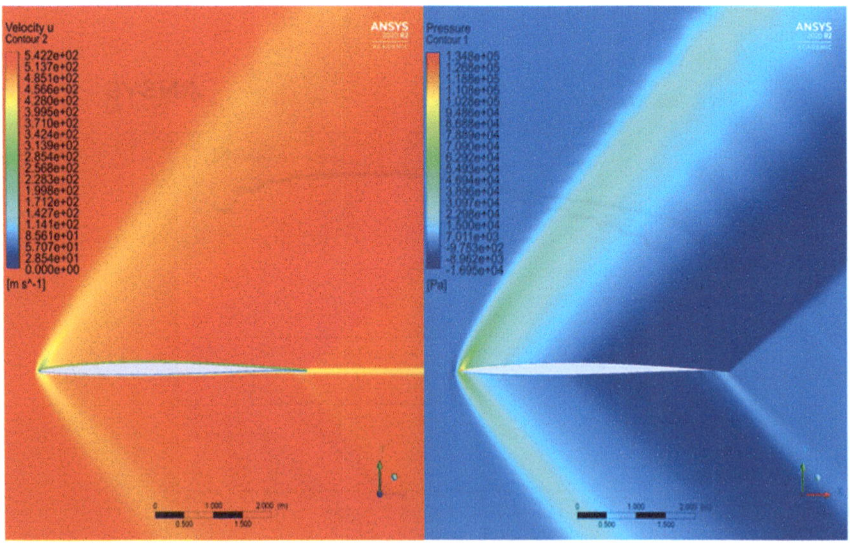

Fig. 9 Pressure and velocity distribution, configuration 0, Ma = 1.5, 0° AoA

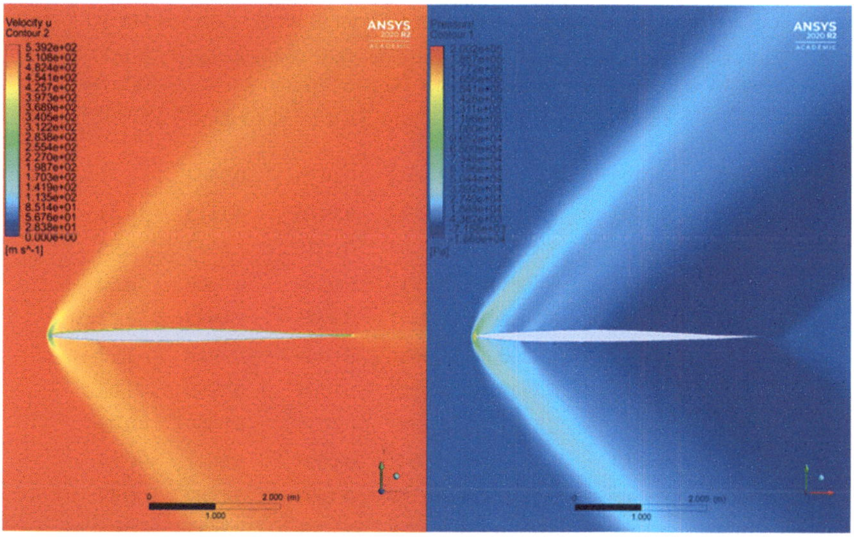

Fig. 10 Pressure and velocity distribution, configuration 1, Ma = 1.5, 0° AoA

Contours of velocity u and total pressure are shown side by side for both configurations, 0 and 1, respectively. The simulation conditions are Ma = 1.5, and 0° AoA. Figures show similar, but different flow around the airfoil. In Fig. 10, it can be seen how a weak shockwave occurs at around 3.5 m from the nose, or at around 78% of chord, just before the rear edge, whereas for configuration 0 there are no shockwaves until the very end of the airfoil.

Next, in Table 3, values obtained by simulating the fluid flow through the discretized domain for both configurations 0 and 1 can be seen. The condition of the flow is Ma = 1.5, and angle of attacks are in range from 0° to 12°.

Furthermore, on Figs. 11 and 12 graphical representations of Table 3 are shown. Figure 11 shows the lift against drag values for both configurations. Black line presents the values obtained for unmodified airfoil (configuration 0), whereas the green line shows values calculated for supersonic configuration (configuration 1) for same conditions Ma = 1.5 and equal range of angles of attack (0° to 12°). Figure 12 presents the lift-to-drag ratio against angle of attack for both configurations. It can be seen how both configurations achieve their maximum at 4° AoA.

In order to validate the results, linear theory was used to calculate the lift coefficients for both configurations at equal conditions, Ma = 1.5 and 0° up to 12° AoA. Figure 13 shows the C_L calculated with CFD and linear theory. The results begin to differ at larger angles of attack, but that is expected since the linear theory has its limitations. Linear theory is limited to thin airfoils and will only give good results for smaller angles of attack, and within the range of smaller angles of attack (AoA < 6°) the results are identical.

Table 3 Lift and drag coefficients predicted by the CFD for 0° till 12° AoA and Ma = 1.5

AoA	Subsonic cruise configuration (conf 0)			Supersonic cruise configuration (conf 1)			Δ in C_L/C_D ratio for configurations 1 and 0 [%]
	C_L	C_D	C_L/C_D	C_L	C_D	C_L/C_D	
0°	−0.0052	0.0195	−0.2684	−0.0044	0.0175	−0.2504	−6.71
1°	**0.0568**	**0.0201**	**2.8207**	**0.0574**	**0.0185**	**3.1115**	**10.31**
2°	0.119	0.023	5.1750	0.1194	0.0216	5.5347	6.95
3°	0.1816	0.0281	6.4688	0.1818	0.0269	6.7600	4.50
4°	**0.2443**	**0.0354**	**6.9066**	**0.2447**	**0.0345**	**7.0942**	**2.72**
5°	0.3085	0.0452	6.8200	0.308	0.0444	6.9410	1.77
6°	0.3719	0.0564	6.5948	0.3724	0.0571	6.5270	−1.03
7°	0.4377	0.0711	6.1579	0.4383	0.0719	6.0969	−0.99
8°	0.505	0.0884	5.7122	0.5044	0.0891	5.6641	−0.84
9°	0.5744	0.1086	5.2890	0.5707	0.1085	5.2587	−0.57
10°	0.6468	0.1318	4.9039	0.6385	0.1304	4.8958	−0.17
11°	0.7239	0.1587	4.5632	0.7102	0.1555	4.5683	0.11
12°	0.8081	0.1894	4.2662	0.7833	0.1835	4.2683	0.05

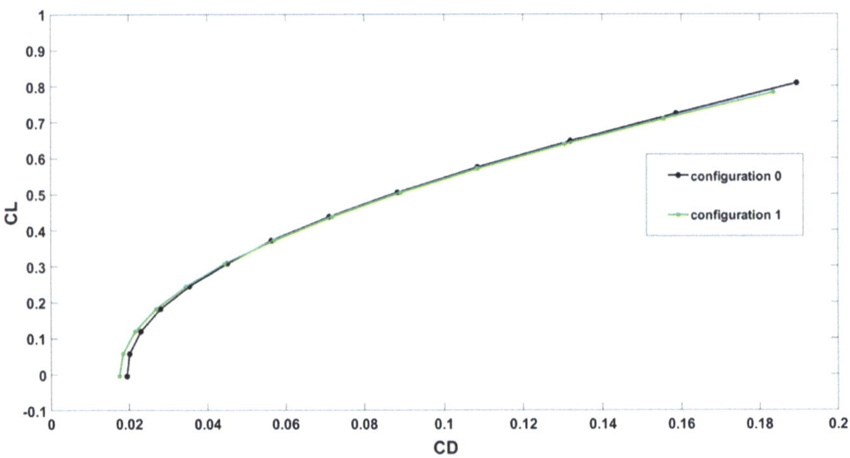

Fig. 11 Lift coefficient against drag coefficient for Ma = 1.5, 0° to 12° AoA, configurations 0 and 1

4 Conclusion

The research goal of this project was to evaluate the impact of the variable geometry airfoil used in the F-16 wings. Even though the geometry can be changed into other various configurations used for different flight regimes, configurations 0 and

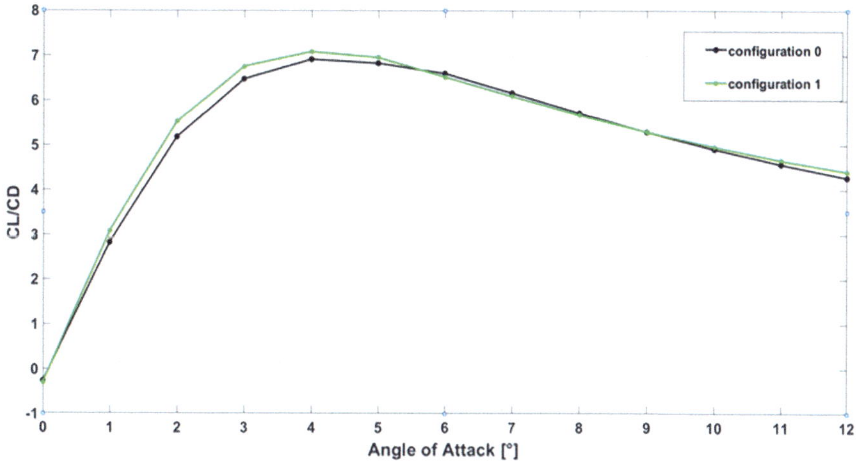

Fig. 12 C_L/C_D against angle of attack, for Ma = 1.5, configurations 0 and 1

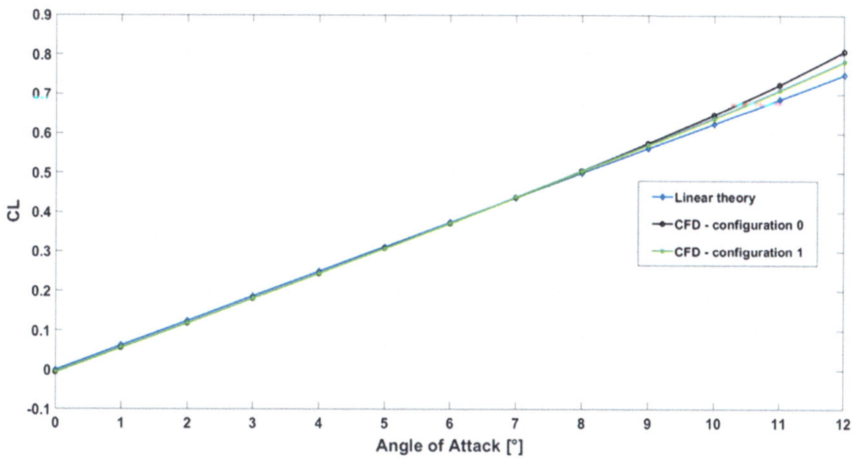

Fig. 13 CFD lift coefficient results validated against supersonic linear theory

1 were chosen for this study. The reason for creating such a complex mechanism for deflection of flaps is beneficial for the performances of the aircraft, but just how much impactful was the goal meant to be discovered. After conducting multiple simulations for different angles of attack at the supersonic regime, the conclusion was that the configuration 1, the configuration made for supersonic flight, gave better C_L/C_D ratio values in the range of 1–5° angle of attack. Notably, at 1° angle of attack the C_L/C_D ratio is over 10% higher for the supersonic configuration. The maximum values for C_L/C_D ratios differ for 0.19 or about 2.75%, at 4° angle of attack, in favour of supersonic configuration. It may not look as much, but for combat aircraft which

fly at supersonic speeds and rely heavily on thrust capacities, such an increase in substantial.

In the horizon of our future research, we see studies to be conducted for determining the optimal angle of deflection for leading and trailing flaps at supersonic regime. Current deflection of 2° was researched; however, by testing additional deflections ranging from 0 to 4°, there might be a better position which could yield even better results. Furthermore, it would be beneficial to research and determine the fuel consumption difference obtained by flying at the finesse angle of attack for both configurations. Additionally, different CFD approaches such as density-based coupled solver, or different turbulence models such as Spalart–Allmaras, or LES Smagorinsky could be used and compared to determine the differences.

References

1. Sogukpinar H (2018) Estimation of supersonic fighter jet airfoil data and low speed aerodynamic analysis of airfoil section at the Mach number 0.15. In: AIP conference proceedings, vol 1935, no 1. AIP Publishing, Turkey, pp 020002–1–020002–4
2. Akgun O, Golcuk AI, Kurtulus DF, Kaynak Ü (2016) Drag analysis of a supersonic fighter aircraf. In: Proceedings of 9th international conference on computational fluid dynamics (ICCFD9), Istanbul, Turkey
3. Hodson JD, Christopherson AP, Deaton JD, Pankoninen AM, Reich GW, Beran PS (2019) Aeroelastic topology optimization of a morphing airfoil in supersonic flow using evolutionary design methods. In: AIAA Scitech 2019 Forum, AIAA. San Diego, USA
4. Hari N, Schetz JA, Kapania RK (2019) Numerical prediction of interference drag of a strut-surface intersection in supersonic flow. In: AIAA Scitech 2019 forum, AIAA. San Diego, USA
5. Askari S, Shojaeefard M, Goudarzi K (2011) Numerical and analytical solution of compressible flow over double wedge and biconvex airfoils. Eng Comput 28(4):441–471
6. Keith T, Cioc S, Jiang H (2019) Spreadsheet computations of a symmetric double wedge airfoil in supersonic flow. Int J Mech Edu 48(3):255–270
7. Manshadi MD, Aghajanian S (2018) Computational aerodynamic optimization of wing-design concept at supersonic conditions by means of the response surface method. J Braz Soc Mech Sci Eng 40(5)
8. Tharkude TS, Li DZ (2018) Simulation study of supersonic natural laminar flow on wing with biconvex airfoil. Int J Eng Res Technol (IJERT) 7(3):366–373
9. Ansys Help Manual (2021) Ansys, Inc. Canonsburg, Pennsylvania, USA
10. Anderson JD (1995) Computational fluid dynamics: the basics with applications. McGraw-Hill, New York

Prediction Tool for Military Flying Operations Planning

Nikola Mostarac, Darko Šumanovac, and Doris Novak

Abstract Military flying operations performed in flexible airspace structures impact commercial air traffic. Analysis of the planning process is performed to define the crucial parts to develop the prediction tool. The process of military flight planning is determined by interactions, conditions, and limitations of the components: flight training syllabus, flight crew, aircraft and associated maintenance, air traffic management, and airspace structures. Military flight planning process model and activity diagram are developed for prediction tool creation. Further system enhancement based on the prediction tool given the real-life operations feedback on specific external factors' influence and internal and mutual interaction is enabled.

Keywords Flexible use of airspace · Air traffic management · Collaborative decision-making · Planning tool

1 Introduction

Military flight management according to [1] defines a group of procedures and measures planned and executed to maintain complete and safe flying, and for analysis of success and measures taken to remove deficiencies. Military flight planning is an integral part of management performed to achieve targeted personnel and unit training capabilities. The higher echelon of command defines these targeted capabilities [1, 2].

The military flight planning process implies particular capability and availability of the air traffic services, intercept guidance services, and radio-navigation assets. Uninterrupted availability of airport services, primary and alternative ones, is mandatory for flight operations. At the same time, the training areas must be available for selected tasks in accordance with the flight training curriculum [1].

A flexible airspace continuum, with all users' requirements fulfilled optimally, presumes coordination bodies. Airspace use planning process, rules, and procedures of civil–military cooperation have to be established. Collaborative decision-making

N. Mostarac (✉) · D. Šumanovac · D. Novak
Faculty of Transport and Traffic Sciences, University of Zagreb, Vukelićeva 4, 10000 Zagreb, Republic of Croatia
e-mail: nikola.mostarac@morh.hr

M. Petrović et al. (eds.), *The Science and Development of Transport—ZIRP 2021*,
https://doi.org/10.1007/978-3-030-97528-9_14

in air traffic flow management is based on the timely exchange of all relevant, reliable, and precise information about air traffic among all relevant airspace users. This approach tends to become a standard [3, 4].

Implementation of Single European Sky regulations does not degrade the sovereignty of member states over their airspace but could potentially lessen the readiness of military forces. It could necessitate additional expenses and could result in the military component's loss of influence. Non-implementation of regulations implies underuse of available resources. It could also create legal implications in case of accidents involving civilian and military participants [5–7].

Civil–military cooperation in the European Community is not uniform; therefore, it limits airspace management and the success of the Single European Sky. The rights and responsibilities of member states in the field of defence are unquestioned by civil–military cooperation. Military operations and training must be protected from the harmful effects of applying common principles and criteria.

Airspace management is conducted at three levels: strategic, pre-tactical, and tactical.

The strategic level (Level 1) plans and recommends the establishment and changes of flexible airspace structures. This level defines and delegates rules of priority for flexible airspace structures. This level is entrusted with coordinating civil and military operational requirements without privilege respecting rules of priority. The joint civil–military procedure performs the strategic level of airspace management [8].

The designated service provider executes level 2 management through a joint civil–military Airspace Management Cell. Ministry of Defence representatives participate in the cell's operations. Pre-tactical airspace management is performed under terms and procedures established at the strategic level [8].

Real-time airspace management (Level 3) is executed by the designated service providers and Ministry of Defence through cooperation and information exchange procedures between civilian and military sectors to activate, deactivate, and re-allocate airspace structures. It is executed throughout the Airspace Management Cell [8, 9].

Prediction Expert System represents a potentially helpful tool for military flying planning effectivity and functionality improvements. It can enhance coordination with civil users' requirements for joint use of airspace. Interconnections and causality of military flight planning process elements must be identified to develop an adequate model for effective tool development [10].

Hazard identification and risk management in the military flight planning process reduce the likelihood of adverse events and increase flight safety. The use of a prediction tool in terms of organizational factors based on objective indicators and knowledge regarding the behaviour of flight participants coupled with an indication of potential conflict solutions contributes to a proactive approach to risk assessment and risk management when planning flight tasks at all levels.

2 Military Flying Planning Elements

The process of military flight planning is determined by interactions, conditions, and limitations of the components: flight training syllabus, flight crew, aircraft and associated maintenance, air traffic management, and airspace structures.

2.1 *Flight Training Syllabus*

Military flight planning is based on a flight training syllabus, which for a particular type of military aircraft determines the goal, tasks, conditions for the start of training, organization, structure, and standards of flight training. The flight training syllabus for personnel and units applied to a certain period, to achieve a defined level of training becomes the flight training plan [11].

The flight training syllabus structures the flight training and represents a template through which the pilot flight training process is conducted as shown in Table 1. The training of pilots of the same type of aircraft does not have to be entirely identical, even within the same country. There is a contemporary tendency to decentralize and bring down the freedom to manage the training syllabus to the squadron level. Throughout history, training has changed significantly, emphasizing extensive and comprehensive.

Structured according to the conditions of executing an individual exercise, the flight training syllabus may be in the form of a sequential flow without branching, as shown in Fig. 1. In such a structured training, each exercise is preceded by one

Table 1 Preview of the flight training syllabus

Designation	Number of flights	Flight duration	Day/night	Formation	Role	Aircraft type	Airspace	Altitude
A1137	1	45	D	P	W	BiS	R	10–20
A1139	2	45	D	P	W	UM	Z	10–20
A1142	3	55	D	P + T	W	BiS	Z	10–20
E1143	1	55	D	P + T	W	UM	R	20–30
A1144	1	45	N	F	W	BiS	Z	20–30
A1146	3	45	N	F	W	UM	Z	30–45
E1149	3	55	N	F + T	W	BiS	Z	30–45

Fig. 1 Flight training syllabus exercise structure, no branches

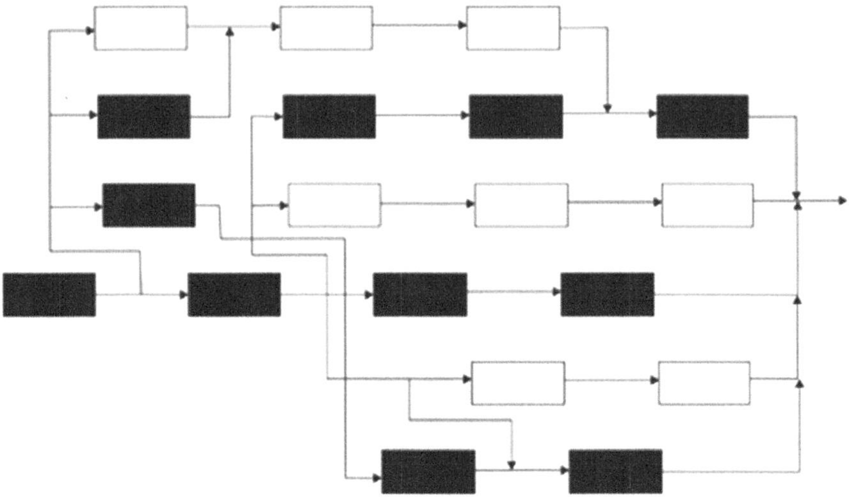

Fig. 2 Flight training syllabus exercise structure, branches

specific exercise. Adverse meteorological conditions and the unavailability of certain training areas can stall the training process.

If more than one previously performed exercise is required for a particular exercise, the flight training flow diagram takes the form shown in Fig. 2. The flow diagram in Fig. 2 shows the basic structure of flight training in which more than one exercise occurs as a prerequisite for specific exercises. Individual branches of the flow chart, the associated exercises may differ in characteristics, such as the required altitude for the implementation of flights, day or night flying. The black rectangles in Fig. 2 represent the exercises performed at night, and the white rectangles describe the exercises performed during the day.

The flight training syllabus prescribes the number of aircraft and the pilots' role in the flight for each exercise. This requirement must be met by the availability of the aircraft, by the continuous availability, and programmed distribution of maintenance in a given time. Depending on the number of aircraft and training requirements, pilots with appropriate training are activated. Training on modern fighter aircraft can be carried out in different proportions of single-seater and two-seater aircraft versions. Flights on two-seater versions enable more challenging and aggressive training due to a kind of fuse in the form of an experienced co-pilot, which reduces the mentality of a single-seat fighter pilot with increased costs and more required pilots [12].

The flight training syllabus also prescribes meteorological conditions and the time of day for performing any individual exercise.

2.2 Pilot Flight Training Within the Squadrons

The required structure and range of flight time required to maintain the level of training are challenging due to the lack of an objective measure of training to determine the minimal and optimal number of flights. Determining the optimal ratio of actual flights to flights on a simulator has no basis in scientific research. However, the benefit of using a simulator in the training process is undoubted. The share of 55% experienced pilots in the total unit pilots' number was recognized as a support for the functioning of the combat squadron. Below that value, the squadron's training was degraded. Crew experience is not based relevantly on flight hours but the tasks or structure of the flight [2, 12].

Squadrons' flight training for targeted capabilities comprises different levels of training. Primary flight training and initial flight training are performed on a particular aircraft type, regardless of the unit's mission. Flight training for maintaining the achieved level of expertise and training are performed after the expiration of the prescribed period frame, in which no flight tasks are performed. Selected individuals receive training for special missions and advanced training such as flight leader or flight instructor training [2, 13].

The status of pilot availability for the implementation of squadron tasks depends on the conducted flight training. Successful completion of the exercises creates the condition for joining the next exercise. Designated exercises are the final exercises for a specified type of training, and their implementation is a prerequisite for the implementation of specified tasks. At this level, the pilot can stagnate and maintain the achieved level of training to implement assigned duties or continue with further flight training in parallel with executing the squadron's dedicated missions and participating in training other pilots in the appropriate role in flight. Pilot training can be conducted to the ultimate level of training without participating in the squadron's dedicated tasks to minimize the period of reaching the full training status.

The planning process is based on the flight training syllabus and the crew training period prescribed by it. According to [1], the training period is "the longest time since the last flight, prescribed by the Flight Training Syllabus, in which a military pilot is considered ready to perform a particular flight task under certain flight conditions on a particular type of aircraft". The length of period and distribution of associated conditioning flights are proportional. Implementing specific exercises and the maintenance of training within the prescribed training period qualify pilots for particular tasks.

The availability of pilots for training and squadron tasks is defined for a specific planning period during which each pilot is assigned to a dedicated squadron task.

2.3 *Aircraft and Associated Maintenance*

The total resource of the aircraft presented as time or the number of operating hours is spent during the whole time of use. Aircraft maintenance must be rigorously executed to achieve availability and reliable use during the entire lifetime. Maintenance can be preventive and corrective. According to the established programme and maintenance plan, preventative maintenance is a planned maintenance service related to the hourly or time resource as timed or periodic inspections. Corrective maintenance is an unplanned maintenance service carried out to eliminate an operational malfunction or general malfunction. Maintenance of military aircraft is carried out according to the maintenance model related to resources or on-condition [14, 15].

The resource-based maintenance model assumes the works on a specific aircraft resource status, while the condition-based maintenance model is based on the condition of the aircraft. Maintenance according to resources is carried out in three stages, and maintenance according to state is carried out as line and base. In both cases, the aircraft is unavailable for flight tasks for a certain period and is subject to preventive work [15].

The total flight time resource of the aircraft is distributed over the lifetime. Resource consumption may vary according to operational needs and the concentration of flight operations within a given time. The available resource is defined for a certain period. The resource of individual aircraft is consumed in such a way as to ensure a stable and even sustainable level of maintenance system workload. Aircraft maintenance operations are planned in correlation with the extent of a defined period of unavailability for use. Maintenance operations are performed at intervals related to the consumption of resources, their duration associated with the complexity of the works [15].

Figure 3 shows an example of structured preventive aircraft maintenance. The shaded fields in the observed period represent the incidence of aircraft's preventive inspections and works. Single-seater versions are marked with BiS and number, and two-seater versions are marked with UM and number. The time distribution of preventative maintenance is achieved by continuous work of the most demanding

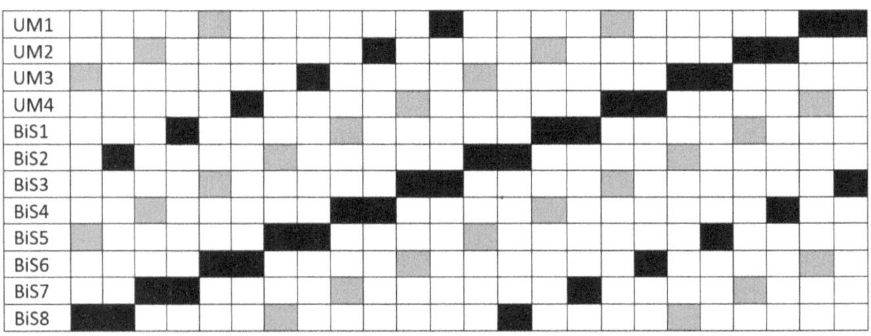

Fig. 3 Aircraft maintenance plan

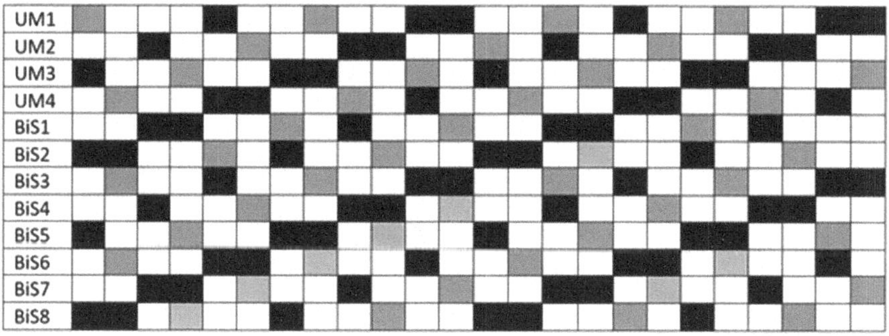

Fig. 4 Aircraft maintenance schedule, higher rate of resource consumption

and extensive maintenance simultaneously on only one aircraft. One workplace is continuously active, where planes appear successively at maintenance. At the same time, one to two planes appear at regular intervals on less demanding and shorter works. With such an organization of resource consumption and preventive maintenance works, a maximum of three and at least two aircraft are under maintenance simultaneously, ensuring the highest availability of 75% and 83%, respectively.

Depending on the reliability and the frequency of failures, additional aircraft appear in the maintenance process. The response to such expected occurrences may be a redirection of human resources from preventive to corrective maintenance, thus creating a shift in preventative maintenance. Suppose corrective maintenance is expected to be more extensive than the available time, idle time between the activation of two and three positions to preserve the resource consumption plan. In that case, it is necessary to have additional human resources and material resources.

Figure 4 shows maintenance during the observed period with an increased rate of resource consumption. It is clear that in addition to increasing the required working positions in a workshop for the implementation of preventive maintenance work, it is not possible to achieve aircraft availability more significantly than 50%. The total available resource has been increased at the expense of the number of available aircraft.

Figure 5 shows maintenance during the observed period with a reduced rate of resource consumption. The number of working positions in a workshop required to implement preventive maintenance works has been reduced. The correctness was achieved from 83 to 91% to the detriment of the total available resource of the aircraft.

2.4 *Air Traffic Management and Airspace Structures*

Sorties are performed in various types of airspaces depending on Flight Training Syllabus sortie specifications. Flights performed inside the airport zone, depending on the number of military aircraft involved, could interfere with civilian traffic.

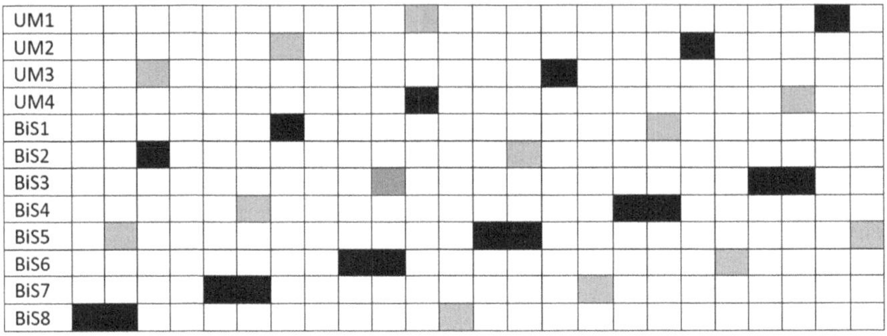

Fig. 5 Aircraft maintenance schedule, lower rate of resource consumption

Following general air traffic rules, flights outside the airport zone do not interfere with civil users. These flights are equally treated as civil ones, and there are no requests for prioritization. The execution of flights following the exercise's specifications defined in the flight training syllabus is related to the volumes of airspace of particular dimensions and at certain altitudes. These zones are recognized as reserved or segregated airspace structures. Zones defined at the strategic level of airspace management are assigned at the pre-tactical and tactical levels. The volume of airspace denied to civilian users varies inactivation of one or more airspace structures. The traffic demand of commercial users does not uniformly burden the altitudes that are dismissed for civil traffic by activating specific structures.

Consequently, the different sizes of airspace structures and the different heights required to perform various exercises create further interference of military aircraft with commercial traffic during tasks. By activating the zone, the planned traffic in the scheduled activation period is redirected to avoid and bypass the designated area. Traffic in zones is distributed in time and represented by the number of flights passing through the zones in correlation with the passage altitudes. Data on traffic in the past period can be used to create a projection of future traffic and as a basis for conducting simulations. Based on the obtained results, it is possible to learn about military and civilian traffic interference.

2.5 *Operational Periods and Restrictions*

Military flight planning is limited by the period for which the planning is carried out and the working hours of the squadron and within the particular legal frame—operating time shifts to enable different tasks and night flight tasks. There are several different types of internal restrictions regarding duty time, flight time, rest time. These restrictions also influence the planning of flying operations and the time prescribed for other portions of instructing process. Airbase support services represent a factor that needs to be taken into consideration for planning. Typical flying operation day's working hours are influenced by: type of flight training, exercise specifics, aircraft preparation peculiarities, apron services, and flight instructors' availability [16].

Military flight planning duration spans periods of the week(s), month(s), or year(s). Depending on the planning period, there are short-term, middle-term, and long-term planning. Each of them benefits directly from the expert system use.

3 Expert System Model for Flight Planning

The military flight planning business' process model is presented in Fig. 6.

Military flight planning is initiated by training orders and other orders issued by the higher echelon of a command containing directions and guidance. Scope and intended capabilities for the planning period are set in plans that include a projection of a path to be followed. Planning process inputs are human potentials, time available, disposable aircraft, and flight time available. Internal restrictions and guidelines for the planning process are set in the flight training syllabus. Planning coordinates internal processes and external factors such as infrastructure availability, training installations availability, and airspace structures status.

The activity diagram in Fig. 7 presents the concept of the planning process's business processes and activities flow.

The planning process begins with the decision to start planning. The pilot's availability, training status, and scheduled duties are analysed. The airplane's flight time available regarding maintenance peculiarities distribution over planning interval is considered to incorporate training priorities properly. These lines of activities result in the pilot's availability and aircraft overviews for the planning period. Depending on flight tasks' priorities, pilots are selected and assigned to aircraft seats available according to their proficiency and training needs. Setting pilots to seats prevents them from being multiple planned in the current sortie line.

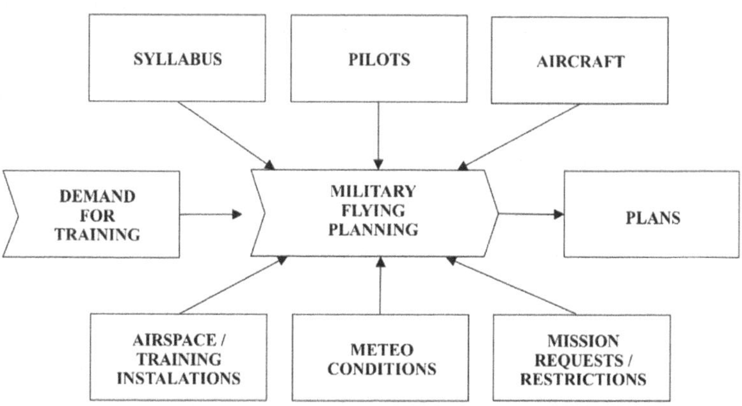

Fig. 6 Military flight planning business' process model

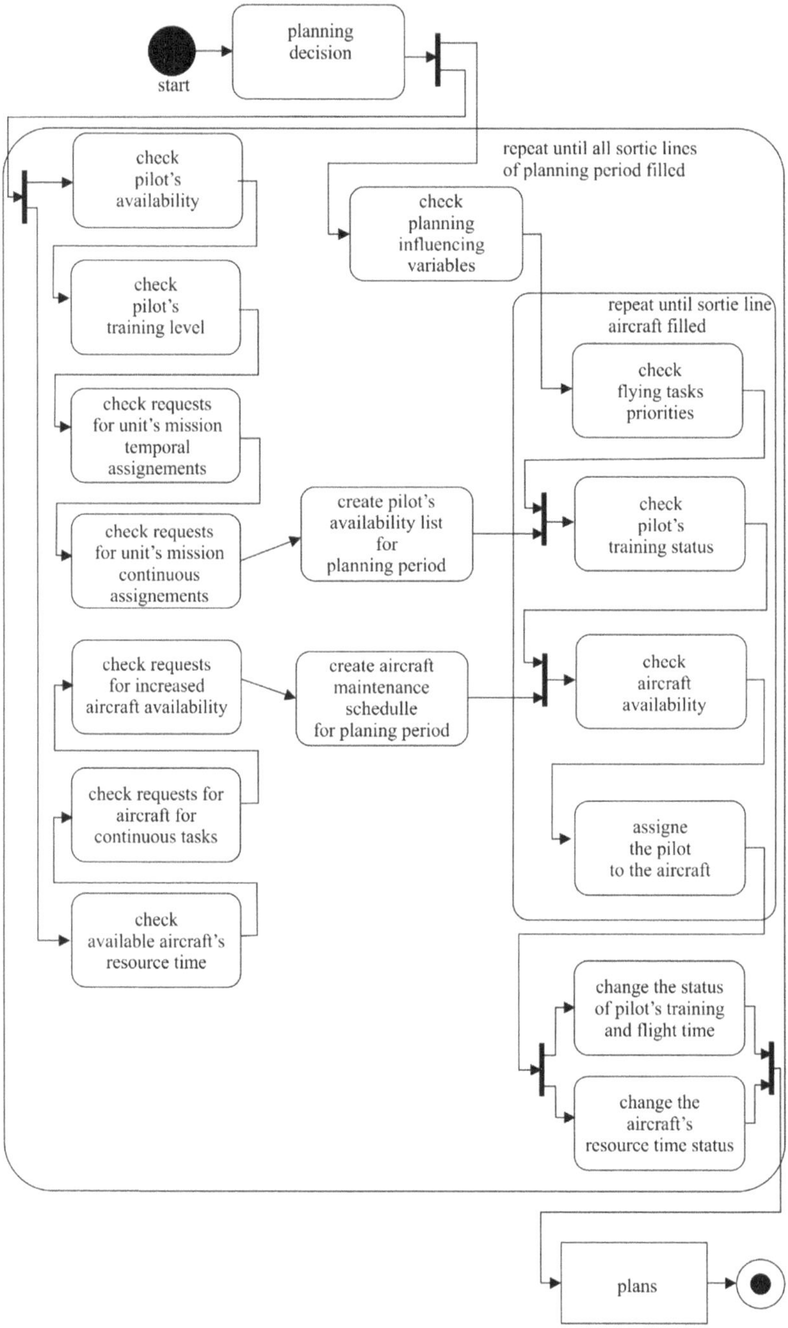

Fig. 7 Military flight planning activity diagram

The basic planning cycle closes once all seats available in one unit sortie line are filled. Individual pilot's training status and aircraft flight time status are updated. Further iterations of the planning process are performed until the end of the planning span is reached.

The activity diagram of the military flight planning process presented in Fig. 7 contains planning influencing variables evaluation. The basic planning cycle in which the knowledge about system functioning is collected is a flight day. Knowledge is stored in a knowledge database cumulatively upgraded to shape and boost the directions and restrictions of the planning process. Typically recognized as outside factors data are meteorological conditions, training infrastructure availability, and other traffic. Certain internal factors such as aircraft availability are systematically controlled and interpreted to be translated to planning parameters. Knowledge regarding training effectiveness and quality is implemented in the curriculum reshaping. The ability to react to deficiencies and predict imperfections by process adjustments indicates the system's quality, resilience, and capability to answer unexpected challenges.

4 Conclusion

The applicability, significance, and value of knowledge evaluated in the planning process and accumulated in the fund of expert knowledge are determined at the planning level of process management. It is conditioned by the specifics of the aircraft and flight unit's mission. Determining the causality of knowledge collected through operation implies its qualification and valuation harmonized with other flight operations management participants.

Co-relation of input parameters and output goals, resources, scenarios, scope, and content of knowledge regarding participants and business process objects implies peculiar specifications of the prediction tool. Further analysis and structuring of the process model flow diagram requires the detailed insight of structures and business processes and represents the object of interest for operational and development institutions.

A certain level of appreciation and adjustment to specifics of airspace users, not to degrade capabilities or upgrade the complexity of operations, represents a challenge that can be answered with the proper act or clear statement of inability to compromise due to operational restrictions. A systematic and collaborative approach of participants is required alongside the clear and precisely defined strategy and scopes. The prediction tool restricted to a single-sided system remains a simple documenting tool with no potential to support the planning process by relevant prediction.

References

1. The Republic of Croatia, Ministry of Defence, Regulation on military airplanes flying (NN 72/2016)
2. US Government Accountability Office (2020) Report to congressional committees ready aircrew program air force actions to address congressionally mandated study on combat aircrew proficiency, US Government Accountability Office, 2020, Available from: https://www.gao.gov/assets/710/707060.pdf. Accessed 25 Mar 2021
3. ICAO (2011) Cir 3300 civil-military cooperation in air traffic management, ICAO
4. ICAO (2006) The convention on international civil aviation, Doc 7300/9, ICAO
5. EU (2004) Regulation (EC) No 549/2004 of the European parliament and of the council of 10 March 2004 laying down the framework for the creation of the single European sky (the framework Regulation).Official J Euro Union
6. EU, Regulation (EC) No 1070/2009 of the European parliament and of the council of 21 October 2009 amending Regulations (EC) No 549/2004, (EC) No 550/2004, (EC) No 551/2004 and (EC) No 552/2004 in order to improve the performance and sustainability of the European aviation system. Official J Euro Union 14 Nov 2009. Accessed 29 Mar 2021
7. Eurocontrol (2019) Guidelines for the implementation of the single European sky by the military, Eurocontrol
8. EU (2005) Commission regulation (EC) No. 2150/2005 of 23 December 2005 laying down common rules for the flexible use of airspace. Official J Euro Union (2005)
9. EU (2019) Commission implementing regulation (EU) 2019/123 of 24 January 2019 laying down detailed rules for the implementation of air traffic management (ATM) network functions and repealing Commission Regulation (EU) No 677/2011. Official J Euro Union
10. Mostarac N, Pavlin S, Modić A (2015) Base for development of prediction tool for civil-military flying operations planning. In: Ščukanec A, Babić D (eds) Proceedings of the international scientific conference. Science and traffic development, 12th May 2015, Zagreb, Croatia, pp 125–134
11. The Republic of Croatia, Ministry of Defence, Regulation on flying service in Croatian armed forces (NN 72/2015)
12. Smith J (2017) Comparative analysis of the USAF F-16 and RAAF [Ph.D. thesis]. Walden University. Cited 31 Mar 2021. Available from: https://scholarworks.waldenu.edu/dissertations/4563/
13. HQ USAF (2010) Air force instruction 11–202 vol 1 flying operations aircrew training, Department of the Air Force USA. Available from: https://static.e-publishing.af.mil/production/1/af_a3/publication/afman11-202v1/afman11-202v1.pdf. Accessed 30 Mar 2021
14. The Republic of Croatia, Ministry of Defence, Regulation on continuous airworthiness and maintenance of military airplanes, (NN 86/13)
15. Kozanidis G, Liberopoulos G, Pitsilkas C (2021) Flight and maintenance planning of military aircraft for maximum fleet availability. Mil Oper Res **15**(1):53–73, 2010. Available from: https://www.researchgate.net/publication/263687210_Flight_and_Maintenance_Planning_of_Military_Aircraft_for_Maximum_Fleet_Availability. Accessed 28 Mar 2021
16. Joković S (2001) Organizacija letačke obuke u pilotskoj školi. Fakultet prometnih znanosti, Zagreb

The manufacturer's authorised representative in the EU is Springer Nature Customer Service Centre GmbH, Europaplatz 3, 69115 Heidelberg, Germany. If you have any concerns regarding our products, please contact ProductSafety@springernature.com

Printed and bound by CPI Group (UK) Ltd, Croydon, CR0 4YY
30/09/2026
02225498-0007